Combinatorics, Computing and Complexity

Mathematics and Its Applications (Chinese Series)

Combinatorics, Computing and Complexity

Edited by

Du Dingzhu

Institute of Applied Mathematics, Academia Sinica,
Beijing, People's Republic of China

and

Hu Guoding

Nankai Mathematics Institute,
Tianjin, People's Republic of China

Kluwer Academic Publishers

DORDRECHT / BOSTON / LONDON

Library of Congress Cataloging in Publication Data

ISBN 7-03-001438-3/O · 302 (外)
ISBN 0-7923-0308-3

Kluwer Academic Publishers incorporates
the publishing programmes of
D. Reidel, Martinus Nijhoff, Dr W. Junk and MTP Press.

Distributors for People's Republic of China:
Science Press, Beijing.

Sold and distributed in the U.S.A. and Canada
by Kluwer Academic Publishers,
101 Philip Drive, Norwell, MA 02061, U.S.A.

In all other countries, sold and distributed
by Kluwer Academic Publishers Group,
P.O. Box 322, 3300 AH Dordrecht, The Netherlands.

Printed in Hong Kong

SERIES EDITOR'S PREFACE

'Et moi, ..., si j'avait su comment en revenir, je n'y serais point allé.'

Jules Verne

The series is divergent; therefore we may be able to do something with it.

O. Heaviside

One service mathematics has rendered the human race. It has put common sense back where it belongs, on the topmost shelf next to the dusty canister labelled 'discarded nonsense'.

Eric T. Bell

Mathematics is a tool for thought. A highly necessary tool in a world where both feedback and nonlinearities abound. Similarly, all kinds of parts of mathematics serve as tools for other parts and for other sciences.

Applying a simple rewriting rule to the quote on the right above one finds such statements as: 'One service topology has rendered mathematical physics ...'; 'One service logic has rendered computer science ...'; 'One service category theory has rendered mathematics ...'. All arguably true. And all statements obtainable this way form part of the raison d'être of this series.

This series, *Mathematics and Its Applications*, started in 1977. Now that over one hundred volumes have appeared it seems opportune to reexamine its scope. At the time I wrote

> "Growing specialization and diversification have brought a host of monographs and textbooks on increasingly specialized topics. However, the 'tree' of knowledge of mathematics and related fields does not grow only by putting forth new branches. It also happens, quite often in fact, that branches which were thought to be completely disparate are suddenly seen to be related. Further, the kind and level of sophistication of mathematics applied in various sciences has changed drastically in recent years: measure theory is used (non-trivially) in regional and theoretical economics; algebraic geometry interacts with physics; the Minkowsky lemma, coding theory and the structure of water meet one another in packing and covering theory; quantum fields, crystal defects and mathematical programming profit from homotopy theory; Lie algebras are relevant to filtering; and prediction and electrical engineering can use Stein spaces. And in addition to this there are such new emerging subdisciplines as 'experimental mathematics', 'CFD', 'completely integrable systems', 'chaos, synergetics and large-scale order', which are almost impossible to fit into the existing classification schemes. They draw upon widely different sections of mathematics."

By and large, all this still applies today. It is still true that at first sight mathematics seems rather fragmented and that to find, see, and exploit the deeper underlying interrelations more effort is needed and so are books that can help mathematicians and scientists do so. Accordingly MIA will continue to try to make such books available.

If anything, the description I gave in 1977 is now an understatement. To the examples of interaction areas one should add string theory where Riemann surfaces, algebraic geometry, modular functions, knots, quantum field theory, Kac-Moody algebras, monstrous moonshine (and more) all come together. And to the examples of things which can be usefully applied let me add the topic 'finite geometry'; a combination of words which sounds like it might not even exist, let alone be applicable. And yet it is being applied: to statistics via designs, to radar/sonar detection arrays (via finite projective planes), and to bus connections of VLSI chips (via difference sets). There seems to be no part of (so-called pure) mathematics that is not in immediate danger of being applied. And, accordingly, the applied mathematician needs to be aware of much more. Besides analysis and numerics, the traditional workhorses, he may need all kinds of combinatorics, algebra, probability, and so on.

In addition, the applied scientist needs to cope increasingly with the nonlinear world and the extra mathematical sophistication that this requires. For that is where the rewards are. Linear models are honest and a bit sad and depressing: proportional efforts and results. It is in the non-linear world that infinitesimal inputs may result in macroscopic outputs (or vice versa). To appreciate what I am hinting at: if electronics were linear we would have no fun with transistors and computers; we would have no TV; in fact you would not be reading these lines.

There is also no safety in ignoring such outlandish things as nonstandard analysis, superspace and anticommuting integration, p-adic and ultrametric space. All three have applications in both electrical engineering and physics. Once, complex numbers were equally outlandish, but they frequently proved the shortest path between 'real' results. Similarly, the first two topics named have already provided a number of 'wormhole' paths. There is no telling where all this is leading - fortunately.

Thus the original scope of the series, which for various (sound) reasons now comprises five subseries: white (Japan), yellow (China), red (USSR), blue (Eastern Europe), and green (everything else), still applies. It has been enlarged a bit to include books treating of the tools from one subdiscipline which are used in others. Thus the series still aims at books dealing with:

- a central concept which plays an important role in several different mathematical and/or scientific specialization areas;
- new applications of the results and ideas from one area of scientific endeavour into another;
- influences which the results, problems and concepts of one field of enquiry have, and have had, on the development of another.

The present volume, the first in the MIA (China) subseries, is concerned with, roughly, the interaction of computing and combinatorics. This includes combinatorial optimization and computational complexity aspects of all kinds of combinatorial problems. This is a field of research that owes its existence largely to the availability of the (large) computing machines. All kinds of new mathematical questions arise from it and, all in all, the field represents a third way in which the big machines are a boon to mathematical research.

A central area of interest can perhaps be stated as follows. Once we (think we) know what is computable and what is not computable, a quantitative theory of computation is needed.

Many and fundamental questions thus arise and this collection of tutorial, survey and other selected papers addresses many of them.

The shortest path between two truths in the real domain passes through the complex domain.

J. Hadamard

La physique ne nous donne pas seulement l'occasion de résoudre des problèmes ... elle nous fait pressentir la solution.

H. Poincaré

Never lend books, for no one ever returns them; the only books I have in my library are books that other folk have lent me.

Anatole France

The function of an expert is not to be more right than other people, but to be wrong for more sophisticated reasons.

David Butler

Amsterdam, March 1989

Michiel Hazewinkel

PREFACE

The papers in this book were presented at the international symposium on combinatorial optimization, held on August 14–26, 1988, in Tianjing and Beijing, China. The symposium was supported by the National Education Committee and the National Science Foundation of China, and held in cooperation with Nankai Mathematics Institute and Institute of Applied Mathematics, Academia Sinica. The organization committee consisted of Yue Minyi (co–chair), Hu Guoding (co–chair), Han Jiye, Du Dingzhu, Yan Jiyi, Li Xiang, Xu Meirui, Xu Shuren and Song Tiantai.

In the symposium, a series of tutorial lectures on structural complexity theory was organized. The first four papers are from such lectures. The rest papers study various topics on combinatorics and computing. They are selected from more than 30 presentations.

We wish to thank all who presented works in the symposium or submitted abstracts for consideration, our colleagues who contributed to the success of the symposium and this book, and the sponsors for their assistance and support.

We are especially grateful to Professor Ron Book for his valuable advice and help in selecting and assembling papers for this book. Without his help, this book could not appear right now.

Finally, We would like to mention that our editorial work was supported in part by the NSF of China.

Table of Contents

WHAT IS STRUCTURAL COMPLEXITY THEORY?[1]

Ronald V. BOOK

Department of Mathematics
University of California, Santa Barbara
Santa Barbara, CA 93106
USA

I. Introduction.

In this paper we present some of the basic results of structural complexity theory, results which have served to motivate much of the activity in the field. In order to put these results into context, it is necessary to develop some perspective.

"It is clear that a viable theory of computation must deal realistically with the quantitative aspects of computing and must develop a general theory which studies the properties of possible measures of the difficulty of computing functions. Such a theory must go beyond the classifications of functions as computable and noncomputable, or elementary and primitive recursive, etc. It must concern itself with computational complexity measures which are defined for all possible computations and which assign a complexity to each computation which terminates. Furthermore, this theory must eventually reflect some aspects of real computing to justify its existence by contributing to the general development of computer science." Hartmanis and Hopcroft [14].

---

[1] This paper is based on lectures given at Peking University and Beijing Computer Institute and at the International Symposium on Combinatorial Optimization, held at the Nankai Institute of Mathematics, Tianjin, People's Republic of China, August 1988. Preparation of this paper was supported in part by the National Science Foundation under grant CCR86-11980.

"A theory of complexity, then, must first define the concept of difficulty (efficiency, complexity). Suppose that we are able to define a notion of 'complexity of algorithm.' One can then discuss the difficulty of a given problem (e.g., computing a function) by the existence or absence of an efficient (relative to the measure of complexity) algorithm for that problem." Borodin [7].

What does it mean to say that one function is more difficult to compute than another? Rabin posed this question in an axiomatic framework (that looks like recursive function theory) that was later developed into an axiomatic theory by Blum. But Cobham [8] approached the same question less formally. Cobham emphasized the need to consider the intrinsic computational difficulty of functions. He wanted a machine-independent or model-independent theory that would allow one to determine the inherent difficulty of functions. As an example, he posed the question of whether multiplication is more difficult than addition and, if so, why?

Cobham developed examples of what he considered to be a machine-independent concept. This concept is now known as the "Ritchie-Cobham" property. A class F of functions has the Ritchie-Cobham property if for any function f, the following are equivalent:

(a) $f \in F$;
(b) the function f is computable by a Turing machine whose running time is bounded above by some function in F;
(c) the function f is computable by a Turing machine that uses an amount of work space that is bounded above by some function in F.

The smallest class known to have the Ritchie-Cobham property is the class of Kalmar-elementary functions (the class E_3 of Grzegorczyk).

Cobham also defined and characterized a class of functions called L: these are functions on the natural numbers that are computable in time bounded by a polynomial in the decimal length of the input. This class was also considered by Edmonds who was interested in determining properties of finite graphs by means of efficient algorithms. Today we consider a function or a problem to be <u>tractable</u> if and only if it can be computed in polynomial time; otherwise it is <u>intractable</u>. One of the important goals of computational complexity theory is to establish a theory of intractability as part of a general *quantitative* theory of complexity.

Prior to 1965, many of the specific problems that were investigated came from the study of the Chomsky hierarchy of formal languages; for example, it was shown that deterministic and nondeterministic pushdown store acceptors have different power so that the class of context-free languages properly includes the class of deterministic context-free languages, and the "LBA problem" was stated: can every context-sensitive language be recognized by a deterministic Turing machine that uses only linear space? However, Hartmanis and Stearns directed attention to the study of complexity classes, specifically, to classes of sequences that could be generated by Turing machines with certain time bounds. This lead to the further investigation of classes specified by Turing machines operating within time or space bounds.

Given a model of computation and a measure of computational difficulty, one would like to be able to determine "trade-offs" with other models and other measures. The two most widely studied measures are time and space. For a wide variety sequential models, changing the model results in at most a polynomial increase in the time or space used. Thus, for many questions it is sufficient to consider just one model and the multitape Turing machine has been considered to be a reasonable candidate. To allow one to focus on the most important problems, we consider Turing machines as acceptors (equivalently, machines compute only characteristic functions), but the reader should feel free to use any familiar programming system that serves as a formalization for the class of computable functions.

Two of the basic classes of interest in structural complexity are referred to as **P** and **PSPACE**. The class **P** of languages recognizable in polynomial time is defined as follows: A set L of strings over a finite alphabet Σ is in the class **P** if and only if there is a deterministic Turing machine M and a polynomial p (of the form $p(n) = n^k$ for some integer $k > 0$) with the property that for all input strings x to M, x is accepted by M if and only if $x \in L$, and for all x, M's computation on x takes at most $p(|x|)$ steps, where $|x|$ denotes the size (length) of x. The class **PSPACE** of languages recognizable in polynomial space is defined as follow: A set L of strings over Σ is in the class **PSPACE** if and only if there is a Turing machine M and a polynomial p (of the form $p(n) = n^k$ for some integer $k > 0$) with the property that for all input strings x to M, x is accepted by M if and only if $x \in L$, and for all x, M's computation on x uses work space bounded by $p(|x|)$.

Notice that P is the class referred to above as the class of tractable problems. It is known that **P** $\subseteq$ **PSPACE** but whether the inclusion is proper is a major open problem.

The mathematical construct of nondeterminism was thought to be of use only in formal language theory until Cook [9] developed the notion of "**NP**-complete problems." The notion of nondeterminism involves the ability to "guess." At this point we will not provide formal definition of the construct; instead we will give a characterization of the class **NP** which is equivalent to the formal definition. A set S of strings over Σ is in the class **NP** if and only if there is a set $A \in$ **P** and a polynomial p with the property that for all strings x, $x \in S$ if and only if there exists a y such that $|y| \leq p(|x|)$ and an encoding $\langle x, y \rangle$ of the pair (x, y) is in A.

The existential quantifier involved in the characterization of **NP** represents the notion of "guessing" $|y|$ choices of bits. A computation of a "nondeterministic time-bounded machine" is considered to consist of a guess of a string of bits and a check of whether the pair consisting of the guess and the input satisfy some relation that can be tested deterministically.

It is easy to see that $\mathbf{P} \subseteq \mathbf{NP}$ and the question of whether the inclusion is proper is one of the most important open problems of theoretical computer science, discrete applied mathematics, and mathematical logic. To see why this is true we must explain the notion of "NP-complete set."

Set A is <u>many-one</u> <u>reducible</u> to set B <u>in polynomial time</u>, written $A \leq_m^P B$, if there is a function f computable in polynomial time with the property that for all x, $x \in A$ if and only if $f(x) \in B$.

A set S is **NP**-<u>complete</u> if $S \in \mathbf{NP}$ and for every set $A \in \mathbf{NP}$, $A \leq_m^P S$.

(The reader may note that the notion of $\leq_m^P$ is simply the idea of "many-one reducibility" from recursive function theory where instead of asking that the function f be computable we demand that it be computable in polynomial time.)

Cook invented the notion of NP-complete set and showed that one specific problem was NP-complete. Consider formulas in propositional (i.e., Boolean) logic in conjunctive normal form (i.e., a conjunction of clauses, each clause being a disjunction of propositional variables or their negations). A formula in propositional logic is <u>satisfiable</u> if there is an assignment of truth values to the propositional variables such that the formula evaluates to "TRUE" under this assignment. Cook proved that the set of conjunctive normal form formulas that are satisfiable, denoted SAT, is NP-complete.

Notice that the problem of checking whether a truth assignment y satisfies a conjunctive normal form formula x is a tractable problem, so that the set of encodings $\langle x, y\rangle$ of pairs such that y is a truth assignment that satisfies x is in **P**. Given x, a "guess" of a truth assignment y that satisfies x corresponds to the use of the existential quantifier. Thus, it is easy to see that SAT is in **NP**. To prove that for every

$A \in$ **NP**, $A \leq_m^P$ SAT, Cook showed that for a nondeterministic

Turing machine M and an input x one can encode M's possible computations on x by a conjunctive normal form formula that is satisfiable if and only if M accepts x. Furthermore, he showed that this can be done in such a way that the size of the formula is bounded by a polynomial in the running time of M; hence, if M runs in polynomial time, then the size of the formula is polynomial in the size of x.

To determine whether a formula in propositional logic is satisfiable, one can use the method of truth-tables. But a formula with n variables requires a truth-table with 2^n rows and so this method requires exponential time. Thus, SAT can be recognized by a deterministic machine that runs in exponential time.

Notice that the class **P** has the property that if $B \in$ **P** and

$A \leq_m^P B$, then $A \in$ **P**. Thus, if any one **NP**-complete problem is

in **P**, then every **NP**-complete problem is in **P** and **P** = **NP**.

Following Cook's work, Karp [16] showed that a wide variety of problems in combinatorial mathematics, operations research, and boolean logic to be **NP**-complete. This caused many other researchers to study the notion of **NP**-completeness and to classify problems as to whether they are in **P** (i.e., tractable) or **NP**-complete problem. It is widely believed that the **NP**-complete problems are intractable, i.e., not in **P**. A long list of **NP**-complete problem was compiled by Garey and Johnson [11] and the list is still growing in size.

A great deal of the motivation for the development of structural complexity theory has come from the study of the **P** =? **NP** problem, of **NP**-complete sets, and of the properties of reducibilities computed by programs with bounded resources ($\leq_m^P$ is only one such reducibility). From this brief background we will describe one view of structural complexity theory.

II. Some Basic Terminology and Notation.

In this section we describe some basic concepts and establish notation.

For a string x, $|x|$ denotes the length of x. For a set S, $\|S\|$ denotes the cardinality of S.

It is assumed that all sets of strings considered are taken over some fixed finite alphabet Σ that contains 0 and 1. The set of all strings over Σ is denoted by Σ^* with the empty string denoted by λ. If $A \subseteq \Sigma^*$, then $\overline{A} = \Sigma^* - A$. If **C** denotes a class of sets, then co-**C** denotes the class $\{\overline{C} \mid C \in \mathbf{C}\}$.

We assume that the reader is familiar with the notion of a multitape Turing machine as a model of computation. We consider machines working as acceptors (i.e., computing characteristic functions); such a machine has a set of <u>accepting</u> states. The basic definition is normally that of a machine that is <u>deterministic</u>, that is, at each step in a computation there is at most one possible action of the machine, i.e., at most one possible next step; in this case, a deterministic machine M accepts an input x if M's (unique) computation on x halts in an accepting state. A machine is <u>nondeterministic</u> if there are some bounded number of possible actions to take at each step. Thus, a nondeterministic machine may have many different computations on a given input; all such computations are represented by a computation tree. A nondeterministic machine M is considered to <u>accept</u> an input x if there is at least one path in M's computation tree on x that ends in a configuration containing an accepting state.

In the midst of a computation, it is sometimes desirable to obtain information from an external source; we refer to such a source as an "oracle" and to machines that utilize an oracle as "oracle machines."

An oracle machine is a multitape Turing machine M with a distinguished work tape, the query tape, and three distinguished states QUERY, YES, and NO. At some step of a computation on an input string x, M may enter the state QUERY. In state QUERY, M transfers into the state YES if the string currently on the query tape is in some oracle set A; otherwise, M transfers into state NO; in either case, the query tape is instantly erased. The set of strings accepted by M relative to the oracle set A is $L(M, A) = \{x \mid$ there is an accepting computation of M on input x when the oracle set is $A\}$. Also, $L(M)$ denotes $L(M, \emptyset)$.

Oracle machines may be deterministic or nondeterministic. An oracle machine may operate within some time bound T, where T is a function of the length of the input string, and the notion of time bound for an oracle machine is just the same as that for an ordinary Turing machine. For any time bound T and oracle set A, $DTIME(T, A) = \{L(M, A) \mid M$ is a deterministic oracle machine that operates within time bound $T\}$ and $NTIME(T, A) = \{L(M, A) \mid M$ is a nondeterministic oracle machine that operates within time bound $T\}$. Of particular interest are the classes $P(A)$ and $NP(A)$ for arbitrary oracle sets A. Here $P(A) = \{L(M, A) \mid M$ is a deterministic oracle machine that operates in time $p(n)$ for some fixed polynomial $p\}$, so that $P(A) = \cup_{k>0} DTIME(n^k, A)$. If $B \in P(A)$, then we say that B is recognized in polynomial time relative to A.

Oracle machines are be used to define Turing reducibilities that are computed within time bounds or space bounds. Thus, set A is Turing reducible to set B in polynomial time, written

$A \leq_T^P B$, if $A \in \mathbf{P}(B)$. It is clear that $\leq_T^P$ is reflexive and transitive. For any class $\mathbf{C}$ of sets, define $\mathbf{P}(\mathbf{C}) = \cup\{\mathbf{P}(A) \mid A \in \mathbf{C}\}$.

The analogous notions are used in defining $\mathbf{NP}(A)$. However, the reducibility $\leq_T^{NP}$ is not transitive.

An oracle machine may operate within some time bound S, where S is a function of the length of the input string; in this case, it is required that the query tape as well as the ordinary work tapes be bounded in length by S. For any space bound S and oracle set A, $\mathrm{DSPACE}(S, A) = \{L(M, A) \mid M$ is a deterministic oracle machine that operates within space bound $S\}$ and $\mathrm{NTIME}(S, A) = \{L(M, A) \mid M$ is a nondeterministic oracle machine that operates within space bound $S\}$. It is known that for appropriate space bounds S, $\mathrm{NSPACE}(S) \subseteq \mathrm{DSPACE}(S^2)$; this result also applies to classes specified by oracle machines. Thus, for arbitrary oracle sets A, $\mathbf{PSPACE}(A) = \{L(M, A) \mid M$ is a deterministic oracle machine that operates in space $p(n)$ for some fixed polynomial $p\} = \{L(M, A) \mid M$ is a nondeterministic oracle machine that operates in space $p(n)$ for some fixed polynomial $p\}$ since $\cup_{k>0}\mathrm{DSPACE}(n^k, A) = \cup_{k>0}\mathrm{NSPACE}(n^k, A)$.

There are some important complexity classes that we can now define.

Let A be a set. Define $\Sigma_0^P(A) = \mathrm{P}(A)$, $\Sigma_1^P(A) = \mathbf{NP}(A)$, and

$\Sigma_{k+1}^P(A) = \mathbf{NP}(\Sigma_k^P(A)) = \{L(M, A) \mid M$ is a nondeterministic oracle machine that runs in polynomial time and $A \in \Sigma_k^P(A)\}$. Then

$\Sigma_0^P(A) \subseteq \Sigma_1^P(A) \subseteq \ldots \subseteq \Sigma_k^P(A) \subseteq \ldots$ is the <u>polynomial-time</u> <u>hierarchy</u> <u>relative</u> <u>to</u> A, and $\Sigma_0^P \subseteq \Sigma_1^P \subseteq \ldots \subseteq \Sigma_k^P \subseteq \ldots$ is the <u>polynomial-time</u> <u>hierarchy</u>. Let $\mathbf{PH}(A) = \cup_{k \geq 0} \Sigma_k^P(A)$ and $\mathbf{PH} = \cup_{k \geq 0} \Sigma_k^P$.

It is clear that for every set A, $\mathbf{PH}(A) \subseteq \mathbf{PSPACE}(A)$. It is not known whether the polynomial-time hierarchy extends beyond $\Sigma_0^P = \mathbf{P}$.

For additional background, the textbooks of Hopcroft and Ullman [15] and of Balcázar, Díaz, and Gabarró [1, 2], and the monograph of Schöning [21] are strongly recommended.

III. A View of Structural Complexity.

Among other topics, structural complexity theory is concerned with the relationships between complexity classes, the structure of complexity classes, the structure of complete sets for complexity classes with respect to different reducibilities, and the various access mechanisms for obtaining information from oracles, that is, the various notions of reducibilities and the corresponding relativizations. The open problems of complexity theory such as **P** =? **NP** suggest the main directions. Here we describe some of the themes that have emerged.

In his doctoral dissertation at the Tokyo Institute of Technology, Watanabe described structural complexity in terms of structural properties. It is often the case that structural properties express the inherent computational difficulty of a problem or of a class of problems. By showing that some specific property is or is not present, it may be possible to describe the difficulty of a problem. For example, **NP**-complete problems are the hardest problems in **NP**. Thus, if a class of problems contains an **NP**-complete problem, then it is considered to be intractable since there is widespread belief and good evidence (but as yet, no proof!) that $\mathbf{P} \neq \mathbf{NP}$, so that problems in $\mathbf{NP} - \mathbf{P}$ are considered to be intractable.

For each structural property we attempt to understand how it witnesses tractability vs. intractability. Structural methods are methods for showing that problems have the corresponding structural properties. Many of the definitions mimic those in recursive function theory but the meanings and interpretations often change. In addition, the results in that theory have served as examples of things one might try to prove in complexity theory; that is, one can examine theorems in recursive function theory and try to prove the analogous result in complexity theory. However, the results often turn out to be very different from the intended analog!

Consider a very simple example. The class of sets that are recognizable by deterministic machines that run in exponential time is denoted by **DEXT**, so $\mathbf{DEXT} = \cup_{k>0}\mathrm{DTIME}(2^{kn})$. Clearly, $\mathbf{P} \subseteq \mathbf{DEXT}$ and a straightforward diagonalization shows that $\mathbf{P} \neq \mathbf{DEXT}$. As noted in the Introduction, $\mathrm{SAT} \in \mathbf{DEXT}$ (in fact, it is likely that the reader can show that his favorite **NP**-complete problem is in **DEXT**). Could it be that $\mathbf{NP} = \mathbf{DEXT}$? The answer is "no" and the proof is based on a simple structural property.

A class **C** is <u>closed under</u> $\leq_m^P$ if $A \leq_m^P B$ and $B \in \mathbf{C}$ imply $A \in \mathbf{C}$. It is easy to see that **NP** is closed under $\leq_m^P$. However, **DEXT** is not closed under $\leq_m^P$ since the closure of **DEXT** under $\leq_m^P$ is $\cup_{k>0}\mathrm{DTIME}(2^{n^k})$ and a diagonalization argument can be used to show that **DEXT** $\neq \cup_{k>0}$DTIME(2^{n^k}).

Thus, **NP** is closed under $\leq_m^P$ but **DEXT** is not, so **NP** $\neq$ **DEXT**. (In fact, it is not known whether **NP** $\subseteq$ **DEXT** or **DEXT** $\subseteq$ **NP**; it is likely to the case that neither inclusion holds.) In this case, closure under $\leq_m^P$ is a structural property that separates **NP** and **DEXT**.

There are a number of examples of structural properties that have proved to be useful in gaining insight into the structure of complexity classes, etc. Some of the most important examples can be described in terms of certain themes: properties of different reducibilities, particularly those that are defined in terms of the basic complexity classes; the density of certain sets, that is, properties that stem from restrictions on the size of sets; and relationships between complete sets with respect to different reducibilities. Many important results can be best understood when considered in terms of these themes.

By reducibilities we mean the binary relations such as $\leq_m^P$ and $\leq_T^P$ defined above. These are special cases of more general notions of many-one and Turing reducibilities from recursive function theory (where instead of demanding that functions be computed within polynomial time it is sufficient to require that the functions be computable, i.e., total recursive). Essentially all of the reducibilities studied in structural complexity theory are defined by considering the more general notions and restricting the computational resources available for their computation.

While $\leq_m^P$ and $\leq_T^P$ are reflexive and transitive, some of the other reducibilities are only reflexive. We will refer to any binary relation that is reflexive as a "reducibility."

By complete sets, we mean more than just "**NP**-complete sets." For a complexity class **C** and a reducibility $\leq_R$, a set B is complete for **C** with respect to $\leq_R$ if $B \in$ **C** and for every $A \in$ **C**, $A \leq_R B$.

By density we mean restrictions on the "growth rate" of sets. In some cases there are properties that stem from restrictions on the density. One class of sets that has received a great deal of attention is the class of "sparse" sets. A set S is (polynomially) sparse if there is a polynomial $p(n) = n^k$ such that for all n, $\|\{x \in S \mid |x| \leq n\}\| \leq p(n)$. Thus, if S is sparse, then its growth rate when compared to the potential lengths of strings is polynomially bounded. A special case is the class of "tally" sets: a set T is a tally set if all of its elements are strings over a one letter alphabet. Thus, a tally set can have at most one string of each length and so is sparse.

We will now consider some examples of how the study of these concepts leads to results about structural complexity.

1. Reducibilities on NP and DEXT.

It is known that $\leq_m^P$ and $\leq_T^P$ differ when considered as binary relations. Do they differ on the class **NP**? We have noted above that **NP** is closed under $\leq_m^P$ but it is an open question whether **NP** is closed under $\leq_T^P$; if **NP** is closed under $\leq_T^P$, then it is closed under complementation and the polynomial-time hierarchy does not extend beyond **NP**. Do $\leq_m^P$ and $\leq_T^P$ differ on the class of **NP**-complete sets? That is, if $B \in$ **NP** and for every set A in NP, $A \leq_T^P B$, is it the case that B is **NP**-complete?

To study such questions Watanabe proposed studying known intractable classes, e.g., **DEXT**. He showed that there exists a set $B \in$ **DEXT** with the property that for every set $A \leq_T^P B$ but there is a set $C \in$ **DEXT** such that $C \not\leq_m^P B$ so that B is $\leq_T^P$-complete for DEXT but not $\leq_m^P$-complete for **DEXT**.

Watanabe has shown that there exists a properly infinite hierarchy of reducibilities computed in polynomial time that between $\leq_m^P$ and $\leq_T^P$, the polynomial time "bounded truth-table" reducibilities, and for each such reducibility there is a set that is complete for **DEXT** with respect to that reducibility but not with respect to any reducibility that lies below it in the hierarchy. Furthermore, none of these complete sets can be sparse; in fact, no sparse set can be complete for **DEXT** with respect to any of the bounded truth-table reducibilities.

2. Restricted relativizations.

Since the idea of reducibilities comes from recursive function theory, one might approach problems such as $\mathrm{P} =? \mathbf{NP}$ by using the methods of elementary recursive function theory. If $\mathrm{P} = \mathbf{NP}$, then one might try to establish this by showing that the computations of a nondeterministic machine recognizing an $\mathbf{NP}$-complete set can be simulated in polynomial time by some deterministic machine. Such a simulation ought to be applicable to oracle machines so that it would follow that for all sets A, $\mathrm{P}(A) = \mathbf{NP}(A)$. On the other hand if $\mathrm{P} \neq \mathbf{NP}$, then one might try to establish this by a simple diagonalization argument. Such a diagonalization ought to be applicable to oracle machine so that it would follow that for all sets A, $\mathrm{P}(A) \neq \mathbf{NP}(A)$. However, Baker, Gill, and Solovay proved that there exist recursive sets A and B such that $\mathrm{P}(A) = \mathbf{NP}(A)$ and $\mathrm{P}(B) \neq \mathbf{NP}(B)$. Hence, these methods of elementary simulation or diagonalization do not appear to be applicable.

In elementary recursive function theory there is a "relativization principle" that can be stated in the following (admittedly, simplistic) way: a statement about sets or collections of sets is true if and only if for every set A, that statement holds for sets or collections of sets relativized to A. But the results of Baker, Gill, and Solovay show that both of the following statements are false: for all sets A, $\mathrm{P}(A) = \mathbf{NP}(A)$ and for all sets B, $\mathrm{P}(B) \neq \mathbf{NP}(B)$. Thus, we conclude that the $\mathrm{P} =? \mathbf{NP}$ problem is not a question of elementary recursive function theory.

The results described in the last paragraphs have stimulated a great deal of interest in reducibilities computed in polynomial time and relativizations of different complexity classes.

If one believes that $\mathrm{P} \neq \mathbf{NP}$, then one might attempt to prove this by studying relativizations of the classes P and $\mathbf{NP}$. The results described above show that further restrictions are necessary. Examining the proof of Baker, Gill, and Solovay showing the existence of a set B such that **P**(B) ≠ **NP**(B), one sees that the size of the set of queries that can be searched is very large. Time bounds serve to bound the number of query strings that can be generated; in n steps a deterministic machine can generate at most n queries while a nondeterministic machine can generate 2^n queries. This observation led to a number of results on "restricted relativizations."

For a machine M, oracle set A, and input x to M, let Q(M, A, x) be defined as {y | there is a computation of M on x relative to A that generates a query to the oracle about y}. For every set A, let $\mathbf{NP}_B(A)$ be the class of all sets L(M, A) where M is a nondeterministic polynomial time-bounded oracle machine and there is a polynomial q that for all x, $\|Q(M, A, x)\| \leq q(|x|)$. Book, Long, and Selman showed that **P** = **NP** if and only if for every set A, $\mathbf{P}(A) = \mathbf{NP}_B(A)$. Thus, if there exists a set A such that $\mathbf{P}(A) \neq \mathbf{NP}_B(A)$, then **P** ≠ **NP**.

While the result described in the last paragraph is concerned with restricting access to an oracle set, there are important results involving oracles of restricted density. Mahaney showed that **P** = **NP** if and only there is a sparse set that is **NP**-complete. Long proved that for some k > 1, if $\Sigma_k^P \neq \Sigma_{k+1}^P$, then for every sparse set S in Σ_k^P, $\mathbf{NP}(S) \neq \Sigma_{k+1}^P$.

The question of whether there are relativization principles that apply to problems of complexity theory naturally arises. The first such principle was developed recently and applies to certain questions about the polynomial time hierarchy. Results originating with Karp and Lipton have been extended and reinterpreted to show that the polynomial-time hierarchy extends to only finitely many levels if and only if there exists a sparse set S such that the polynomial-time hierarchy relative to S extends to only finitely many levels if and only if for all sparse sets S the polynomial-time hierarchy relative to S extends to only finitely many levels.

3. Lowness and Kolmogorov complexity.

How does one measure the amount of information encoded in a set of strings? One method of doing this is to identify this notion with the inherent computational complexity of the characteristic function of the set; for example, a set that can be recognized in exponential time but not polynomial time is considered to encode more information than a set that can be recognized in polynomial time. A second method of doing this is to consider a function that bounds the (generalized) Kolmogorov complexity of the strings in the set. Then a set with large Kolmogorov complexity is considered to encode more information than a set with small Kolmogorov complexity. A third method is to consider the set as an oracle set and apply operators that correspond to resource-bounded reducibilities; for example, if $NP(A) = \Sigma_2^P$ but $NP(B) = \Sigma_3^P$, then B is considered to encode more information than A. Still another method is to consider the notions of "highness" and "lowness." Exploration of potential relationships between these methods has yielded interesting results.

Set $A \in \mathbf{NP}$ is <u>low</u> if there exists an i such that $\Sigma_i^P(A) \subseteq \Sigma_i^P$; the collection of sets $A \in \mathbf{NP}$ such that $\Sigma_i^P(A) \subseteq \Sigma_i^P$ is denoted L_i. It is easy to see that for each i, $L_i \subseteq L_{i+1}$ and the structure $L_0 \subseteq L_1 \subseteq \ldots$ is called the low hierarchy within NP.

Set $A \in \mathbf{NP}$ is high if there exists a j such that $\Sigma^P_{j+1} \subseteq \Sigma^P_j(A)$; the collection of sets $A \in \mathbf{NP}$ such that $\Sigma^P_{j+1} \subseteq \Sigma^P_j(A)$ is denoted H_j. It is easy to see that for each i, $H_i \subseteq H_{i+1}$ and the structure $H_0 \subseteq H_1 \subseteq \ldots$ is called the high hierarchy within NP.

There exists a set in **NP** that is both high and low if and only if the polynomial-time hierarchy extends to only finitely many levels. There exists a set in **NP** that is neither high nor low if and only if the polynomial-time hierarchy extends to infinitely many levels. Every **NP**-complete set is in H_0 while every set in **NP** $\cap$ co-**NP** is in L_1.

The notion of lowness has been extended to sets in the polynomial-time hierarchy. In general, lowness with respect to the classes Σ^P_i, $i > 0$, may be interpreted as setting an upper bound on the amount of information that can be encoded in the set, that is, a low set in the polynomial-time hierarchy has the power of only a bounded number of alternating quantifiers or, equivalently, a bounded number of applications of the **NP**( )-operator. If S is a low set in the polynomial-time hierarchy and $\mathbf{NP} \subseteq \mathrm{P}(S)$, the the polynomial-time hierarchy extends to only finitely many levels - the level depends on the "degree of lowness" of S.

The idea of the "Kolmogorov complexity" of finite strings provides a definition of the notion of the "degree of randomness" of a string. Informally, the Kolmogorov complexity of a finite string is the length of the shortest program that will generate the string; intuitively, it is a measure of the amount of information that the string contains. A string is considered to be "random" if the length of the shortest problem that generates the string is the same as that of the string itself. This concept has been studied extensively and has found many applications in computer science. A modification of the original idea has also been developed: consider not only the length of a program but also, and simultaneously, the running time of the program. One might consider the notion of "a generalized, two-parameter Kolmogorov complexity measure for finite strings which measures how far and how fast a string can be compressed": given a universal Turing machine U and functions G and g, a string x is in the generalized Kolmogorov class $K_U[g(|x|), G(|x|)]$ if there is a string y of length at most $g(|x|)$ with the property that U will generate x on input y in at most $G(|x|)$ steps. Set A has <u>small generalized Kolmogorov complexity</u> if there exist constants c and k such that for every $x \in A$, x is in $K[c \cdot \log|x|, |x|^k]$. The class of sets with small generalized Kolmogorov complexity is precisely the class of sets that are polynomially "isomorphic" to tally sets.

For every set of strings A over a finite alphabet, there is a family $\{C_n \mid n \geq 0\}$ of circuits with the property that for every n and every string x with $|x| = n$, C_n yields output 1 on input x if and only if $x \in A$. Set A has <u>polynomial-size circuits</u> if there exists a polynomial p such that for all n, the size of C_n is bounded above by p(n).

It is known that a set A has polynomial-size circuits if and only if there is a sparse set S such that $A \in P(S)$ if and only if there is a tally set T such that $A \in P(T)$.

Set A has <u>self-producible circuits</u> if there exists a set $B \in P$ and a function $f: \{0\}^* \rightarrow \Sigma^*$ with the properties that (i) for all x, $x \in A$ if and only if $\langle x, f(0^{|x|})\rangle \in B$, and (ii) f can be computed in polynomial time relative to an oracle for the set A.

It is known that a set A has self-producible circuits if and only if there is a tally set T such that $A \leq_T^P T$ and $T \leq_T^P A$. In addition, it is known that the sets with small generalized Kolmogorov complexity form a proper subclass of the class of sets with self-producible circuits which in turn form a proper subclass of the class of sets with polynomial-size circuits. Each of these three classes can be specified in terms of tally sets and, hence, consists of sets that are low. This is very natural since all of the information encoded in a string over a letter alphabet is represented by its length.

4. Separating and Collapsing Hierarchies.

The question of whether the polynomial-time hierarchy extends to infinitely many levels is a major open problem of complexity theory. A hierarchy is said to "collapse" if it extends to only finitely many levels. While there are no results indicating that the polynomial-time hierarchy does or does not collapse, there are such results about the relativizations of the hierarchy, and while there are no results indicating that the union **PH** of the polynomial-time hierarchy is or is not equal to **PSPACE**, there are such results about the relativizations of PH and of PSPACE. Although a collapsing result or a separating result about relativizations of the hierarchy does not directly imply any solution to the unrelativized problem, the techniques developed for the proofs of these results provide some insight into the difficulty of the original problems; in addition, some of these proof techniques are of interest in their own rights.

Proofs involving the construction of oracles that separate or collapse hierarchies often depend on complex counting arguments. Recently, a new technique was introduced by Yao and by Hastad which has yielded lower bound arguments that have been used to show that there exist relativizations of the polynomial-time hierarchy that extend to any desired level. This technique is based on the representation of the computation tree of an oracle machine as a circuit and thus allows for the construction of circuit computation trees which are more structured than those of arbitrary oracle machines.

The following results have been obtained by this method:

a. There exists a set A such that $\mathbf{PH}(A) \neq \mathbf{PSPACE}(A)$.

b. For every k, there exists a set B_k such that $\Sigma_{k-1}^P(B_k) \neq \Sigma_k^P(B_k)$ but $\Sigma_k^P(B_k) = \mathbf{PH}(B_k)$ and $\mathbf{PH}(B_k) = \mathbf{PSPACE}(B_k)$.

c. For every k, there exists a set C_k such that $\Sigma_{k-1}^P(C_k) \neq \Sigma_k^P(C_k)$ but $\Sigma_k^P(C_k) = \mathbf{PH}(C_k)$ while $\mathbf{PH}(C_k) \neq \mathbf{PSPACE}(C_k)$.

d. There exists a set D such that for every k, $\Sigma_{k-1}^P(D) \neq \Sigma_k^P(D)$.

In case d., the polynomial-time hierarchy relative to D extends to infinitely many levels and, hence, $\mathbf{PH}(D) \neq \mathbf{PSPACE}(D)$ (this follows from the fact that there is a set that is $\leq_m^P$-complete for $\mathbf{PSPACE}(D)$ and for every k, $\Sigma_k^P(D)$ is closed under $\leq_m^P$).

These results show that for any combination of possible conditions regarding the relationship between the polynomial-time hierarchy and **PSPACE**, there is a relativization that realizes that possibility.

The notions of lowness and highness in **NP** also give rise to the possibilities of hierarchies within **NP**. It is not known whether either the low hierarchy within **NP** or the high hierarchy within **NP** is infinite. However, the same methods used to prove the relativization results about the polynomial-time hierarchy mentioned above have been used to show that there exists a set A such that both the low hierarchy within **NP**(A) and the high hierarchy within **NP**(A) are infinite. In addition, it has been shown for every k there exists a set B_k such that both the low hierarchy within **NP**(B_k) and the high hierarchy within **NP**(B_k) collapse to level k.

There are many other interesting and important results regarding potentially infinite hierarchies and their relativizations; it is an important topic in structural complexity. Any quantitative theory of complexity that is sufficiently robust as to establish a theory of intractability must provide the tools necessary to develop an understanding of the nature of relativized computation in the presence of resource bounds. In addition, it must take into account the various separating and collapsing results that arise when one considers relativizations and provide an explanation, *through general principles* , of when and how such separation and collapsing can take place.

5. Properties of Complete Sets.

The structure of complete sets with respect to the various types of reducibilities is a topic that receives a great deal of attention. One aspect of this is the "isomorphism conjecture" of Berman and Hartmanis.

Sets A and B are p-<u>isomorphic</u> if $A \leq_m^P B$ is witnessed by a function f that is one-to-one and onto and f^{-1} is computable in polynomial time.

Berman and Hartmanis conjectured that all **NP**-complete sets are p-isomorphic. Notice that if all **NP**-complete sets are p-isomorphic, then $P \neq NP$ since otherwise every nonempty finite set is **NP**-complete but no finite set can be p-isomorphic to any infinite set. This conjecture has not been settled. However, there is evidence that the conjecture does not hold.

For any reducibility $\leq_R$, sets A and B are $\leq_R$-<u>equivalent</u>, denoted $A \equiv_R B$, if $A \leq_R B$ and $B \leq_R A$.

Thus, the conjecture of Berman and Hartmanis is that the collection of **NP**-complete sets forms a single $\equiv_m$ equivalence class.

Following the program of Watanabe described in 1. above, one might ask whether all sets that are $\leq_m^P$-complete for **DEXT** are p-isomorphic. Again, this is not known but there results that provide a good deal of information. For example, all sets that are $\leq_m^P$-complete for **DEXT** are equivalent under $\leq_m^P$-reductions that are one-to-one and length-increasing.

A function is *one-way* if it is one-to-one and computable in polynomial time but its inverse is not computable in polynomial time. The study of one-way functions is important in cryptography where it is necessary to have functions that are easy to compute while the corresponding decryption functions are difficult to compute. It turns out that the study of one-way functions also plays a role in the study of properties of complete sets.

It is known that if one-way functions do not exist, then the $\leq_m^P$-complete sets for **DEXT** are p-isomorphic. If one-way functions do exist, then there exist sets A and B that are equivalent under one-to-one length-increasing $\leq_m^P$-reductions, that are both complete for **DEXT** with respect to "polynomial-time truth-table" reducibility, but are not p-isomorphic.

Consider nondeterministic polynomial time-bounded machines that have the property that for each input accepted there is a unique accepting computation. The class of sets recognized by such machines is denoted by UP.

It is known that one-way functions exist if and only if $P \neq$ UP. (Of course, $P \neq$ UP implies $P \neq$ NP.) It is also known that the notion of p-isomorphism is precisely the same as that of equivalence under one-to-one, length-increasing $\leq_m^P$-reductions if and only if P = UP.

Density considerations also play a role here since if for every $A \in$ UP there exists a sparse set S such that $A \leq_m^P S$, then one-way functions do not exist and P = UP.

6. Complexity Cores.

If a set is intractable, then this cannot be the case because there are only a finite number of candidates whose membership in the set is difficult to decide. The complexity classes under discussion are sufficiently robust as to be invariant under finite variation - Turing machines or programs can be "patched" by finite tables.

Consider a recursive set A that is not in **P**. Since A is recursive, there is an algorithm that recognizes A, i.e., that computes A's characteristic function. Since $A \notin$ **P** for every algorithm α that recognizes A and for every polynomial p, there exist infinitely many inputs x such that α's running time on x exceeds $p(|x|)$. In an entirely different context Lynch developed a remarkable result. If $A \notin$ **P**, then there exists an infinite recursive set X such that for every algorithm α that recognizes A and every polynomial p, α's running time on x exceeds $p(|x|)$ on all but finitely many $x \in X$. The set X is a _polynomial complexity core_ of A. Thus, a polynomial complexity core is a set of hard instances for candidates for membership in A and every algorithm that recognizes A must take more than any polynomial amount of time to decide membership in A for all but finitely many elements in this core.

If a recursive set is not in **P**, then one can show that it has an infinite subset that is itself a complexity core; such a core is called _proper_. This means that a proper complexity core of a set is a subset that is intrinsically difficult to recognize, so that every recursive not in **P** has such a subset.

An intractable set may have an infinite subset that is itself tractable; such a set is called a **P**-_approximation_. If a set has no **P**-approximation, then it is **P**-_immune_. If an intractable set has a **P**-approximation, then one may consider the relationship between such a **P**-approximation and a proper complexity core for that set. Set A is _almost_ **P**-_immune_ if there exist disjoint sets B and C such that $A = B \cup C$, $B \varepsilon$ **P**, and C is **P**-immune.

A set is almost **P**-immune if and only if it has a maximal proper complexity core if and only if it has a maximal **P**-approximation, where maximality is with respect to inclusion modulo finite differences in the lattice of subsets of Σ^* specified by that partial order. Thus, an intractable set has a nontrivial maximal subset that is intrinsically difficult to recognize if and only if it has a nontrivial maximal subset that is intrinsically easy to recognize.

The answer to the question of whether an intractable set is almost P-immune determines the structure of the collection of all proper polynomial complexity cores of that set. This collection forms an ideal, hence a sublattice, of the lattice of subsets of Σ^* described in the last paragraph. The structure of this lattice can only be of two types for any intractable set, one type if the set is almost P-immune and another type if it is not.

A polynomial complexity core of an intractable set is itself P-immune so is itself intractable. What can be said about its density? If an intractable set is sparse, then any subset is sparse so that every proper core is sparse. If an intractable set is not sparse, then must it have nonsparse polynomial complexity cores or nonsparse proper cores? Questions of this type have been thoroughly studied.

The basic idea of polynomial complexity core has been generalized to classes other than P. Du provided a conceptually simple generalization of the notion of polynomial complexity core by emphasizing properties of proper cores. For any class **C** of sets (of strings over a finite alphabet) and any set A, let $\mathbf{C}_A$ denote the collection of subsets of A that are in **C**. A set H is a <u>hard core for</u> A <u>with respect to</u> **C** if for every C in $\mathbf{C}_A \cup \mathbf{C}_{\overline{A}}$, $C \cap H$ is finite; H is <u>proper</u> if $H \subseteq A$.

Similar to the basic result about polynomial complexity cores, there is the following result about hard cores: If **C** is a countable class that is closed under finite union and finite variation, then any set A not in **C** has a proper hard core with respect to **C**. Many of the properties of polynomial complexity cores having to do with complexity, density, and structure can be established for proper hard cores with respect to such classes. In one sense the fact that these properties can be so translated provides evidence that the results on polynomial cores do not depend on deep properties of **P** but rather on fairly general properties of many widely studied complexity classes.

With only a few exceptions the references given for the material described in this section are secondary sources, i.e., textbooks, monographs, or survey, tutorial, or "overview" papers.

For information about the comparison of properties of reducibilities on NP and DEXT, see Watanabe's dissertation [24] and his paper [25]. Recently, Book presented a tutorial paper [5] on restricted relativizations; the paper by Mahaney [20] contains a great deal of information about sets reducible to sparse sets. The notion of "lowness" was introduced in computational complexity theory by Schöning; see his monograph [21]. Li and Vitanyi have written an important survey [19] of Kolmogorov complexity. The paper by Tang [23] and the paper by Book [4] cover other topics described in part 3. The paper by Ko [17] provides a description and analysis of the proof techniques that have yielded the results described in part 4. Properties of complete sets have been studied by many researchers and connections with questions of cryptography are imporant; for information on this topic, see the textbook by Balcázar, Díaz, and Gabarró [2], and the papers by Ko, Long, and Du [18], by Grollman and Selman [12], by Mahaney [20], by Selman [22], and by Watanabe [26]. The material in part 6 was surveyed by Book, Du, and Russo [6] and the basic results on generalized complexity cores were developed by Du [10].

Notes on the References

The textbook by Balcázar, Díaz, and Gabarró [1] contains material on which structural complexity theory is based, and the forthcoming second volume [2] will be extremely valuable for those interested in the topics described here. The monograph by Schöning [21] is very useful. The collection [13] of papers edited by Hartmanis presents some material that lies at the heart of structural complexity theory and other material that is closely related. The collection [3] of papers edited by Book presents material that is important for structural complexity theory.

Current research in structural complexity theory is reported in a variety of journals devoted to theoretical computer science and in the Proceedings of the Annual IEEE Conference on Structure in Complexity Theory, sponsered by the IEEE Computer Society. The proceedings of various conferences contain papers on the topic. Several times per year, the EATCS Bulletin contains a "column" on structural complexity theory.

References

1. Balcázar, J., Díaz, J., and Gabarró, J., Structural Complexity I, Springer-Verlag, 1988.
2. Balcázar, J., Díaz, J., and Gabarró, J., Structural Complexity II, Springer-Verlag, 1990, to appear.
3. Book, R. (ed.), Studies in Complexity Theory, Research Notes in Theoretical Computer Science, Pitman Publ., 1986.
4. Book, R., Sparse sets, tally sets, and polynomial reducibilities, Math. Found. of Computer Sci. 1988, Lecture Notes in Computer Science (324), 1988, 1-13.
5. Book, R., Restricted relativizations of complexity classes, in [13].
6. Book, R., Du, D.-Z., and Russo, D., On polynomial and generalized complexity cores, Proc. 3rd IEEE Conf. on Structure in Complexity Theory, 1988, 236-250.
7. Borodin, A., Computational complexity: theory and practice, in A. V. Aho (ed.), Currents in the Theory of Computing, Prentice-Hall (1973), 35-89.

8. Cobham, A., The intrinsic computational difficult of functions, Proc. 1964 Congress for Logic, Methodology, and Philosophy of Science, North-Holland (1964), Amsterdam, 24-30.
9. Cook, S., The complexity of theorem-proving procedures, Proc. 3rd ACM Symp. Theory of Computing (1971), 151-158.
10. Du, D.-Z., Generalized Complexity Cores and Levelability of Intractable Sets, Doctoral Dissertation, University of California, Santa Barbara, 1985.
11. Garey, M., and Johnson, D., Computers and Intractability - A Guide to the theory of NP-completeness, Freeman (1979).
12. Grollman, J., and Selman, A., Complexity measures for public-key cryptosystems, SIAM J. Computing 17 (1988), 309-335.
13. Hartmanis, J. (ed.), Computational Complexity Theory, Proc. of a Symp. in Appl. Math., American Math. Soc., 1989, to appear.
14. Hartmanis, J., and Hopcroft, J., An overview of the theory of computational complexity, J. Assoc. Computing Machinery 18 (1970), 444-475.
15. Hopcroft, J., and Ullman, J., Introduction to Automata Theory, Languages, and Computation, Addison-Wesley (1979).
16. Karp, R., Reducibility among combinatorial problems, in R. Miller and J. Thatcher (eds.), Complexity of Computer Computations, Plenum Press (1972), 85-103.
17. Ko, K.-I, Constructing oracles by lower bound techniques for circuits, to appear.
18. Ko, K.-I, Long, T., and Du, D.-Z., On one-way functions and polynomial-time isomorphisms, Theoret. Computer Sci. 47 (1986), 263-276.
19. Li, M., and Vitanyi, P., Two decades of applied Kolmogorov complexity, Proc. 3rd IEEE Conference on Structure in Complexity Theory, 1988, 80-101.
20. Mahaney, S., Sparse sets and reducibilities, in [3].
21. Schöning, U., Complexity and Structure, Lecture Notes in Computer Science (211), 1985.
22. Selman, A., Complexity issues in cryptography, in [13].
23. Tang, S., Randomness, tally sets, and complexity classes, to appear.
24. Watanabe, O., On the Structure of Intractable Complexity Classes, Doctoral Dissertation, Tokyo Institute of Technology, 1987.
25. Watanabe, O., A comparison of polynomial time completeness notions, Theoret. Computer Sci. 24 (1987), 249-265.
26. Watanabe, O., On one-way functions, to appear.

Constructing Oracles by Lower Bound Techniques for Circuits [1]

Ker-I Ko
Department of Computer Science
State University of New York at Stony Brook
Stony Brook, NY 11794

Separating or collapsing complexity hierarchies has always been one of the most important problems in complexity theory. For most interesting hierarchies, however, we have been so far unable to either separate them or collapse them. Among these unsolvable questions, whether P equals NP is perhaps the most famous one. In view of the fundamental difficulty of these questions, a less interesting but more realistic alternative is to consider the question in the relativized form. Although a separating or collapsing result in the relativized form does not imply directly any solution to the original unrelativized question, it is hoped that from such results we do gain more insight into the original questions and develop new proof techniques toward their solutions. Recent investigation in the theory of relativization shows some interesting progress in this direction. In particular, some separating results on the relativized polynomial hierarchy have been found using the lower bound results on constant-depth circuits [Yao, 1985; Hastad, 1986, 1987]. This new proof technique turns out to be very powerful, capable of even collapsing the same hierarchy (using, of course, different oracles) [Ko, 1989].

In this paper, we survey recent separating and collapsing results on several complexity hierarchies, including the polynomial hierarchy, the probabilistic polynomial hierarchy, the bounded Arthur-Merlin hierarchy, the generalized Arthur-Merlin hierarchy (or, the interactive proof systems), and the low hierarchy in NP. All these results rely on the newly developed lower bound results on constant-depth circuits.

[1] This paper is based on a lecture presented in the International Symposium on Combinatorial Optimization, held at Nankai Institute of Mathematics in Tienjing, People's Republic of China, August, 1988; research supported in part by the NSF Grant CCR-8801575.

We show how these new combinatorial proof techniques are combined with the classical recursion-theoretic proof techniques to construct the desirable oracles. Our focus is on how the diagonalization and the encoding requirements can be set up without interference to each other and how such complicated requirements can be simplified using lower bound results on constant-depth circuits.

In Section 1, we give the formal definitions of the relativized complexity hierarchies mentioned above. We avoid the definitions of different machine models but use the polynomial length-bounded quantifiers to give simpler definitions. This allows us to define easily, in Section 2, the related circuits for different complexity hierarchies. In Section 3, the basic proof technique of diagonalization is introduced and a simple example is given to show how oracles for separation results are constructed in this standard setting. Section 4 shows how the lower bound results on circuits can be combined with the diagonalization technique to separate various complexity hierarchies. In Section 5, the proof technique of encoding is introduced. Examples are given to show that to collapse hierarchies to a particular level requires both the diagonalization and the encoding techniques, together with more general lower bound results on circuits. These results on some hierarchies are given in Section 6. Section 7 deals with less familiar hierarchies as the applications of the standard proof techniques. This includes the generalized Arthur-Merlin hierarchy and the low hierarchy in *NP*. The last section lists some open questions in the theory of relativization of hierarchies related to our approach.

Notation. We will deal with strings over alphabet $\{0,1\}$. For each string x, let $|x|$ denote its length, and for each finite set A, let $\|A\|$ denote its cardinality. We write $A^{=n}$, $A^{<n}$, and $A^{\leq n}$ to denote the subsets $\{x \in A \mid |x| = n\}$, $\{x \in A \mid |x| < n\}$ and $\{x \in A \mid |x| \leq n\}$, respectively, while $\{0,1\}^n$ denotes all strings over $\{0,1\}$ of length n. We let $\langle, \cdots, \rangle$ be a fixed one-to-one pairing function from $\cup_{n\geq 1}(\{0,1\}^*)^n$ to $\{0,1\}^*$ such that $|x_i| < |\langle x_1, \cdots, x_n\rangle|$ for all $i \leq n$ if $n > 1$. All of our use of log is the logarithm function of base 2.

1. Relativized Complexity Classes

We begin with the formal definitions of relativized complexity classes. We assume that the reader is familiar with ordinary deterministic Turing machines (TMs) and the complexity classes P, NP and $PSPACE$.[2] All the relativized complexity classes to be considered here, except the class $PSPACE(A)$, can be defined in terms of polynomial length-bounded quantifiers over polynomial-time predicates. So, we will only define the deterministic oracle machine and avoid specific machine structures of nondeterministic, probabilistic and alternating machines.

A (deterministic) oracle TM is an ordinary TM equipped with an extra tape, called the *query* tape, and three extra states, called the *query* state, the *yes* state and the *no* state. The query tape is a write-only tape to be used to communicate with the oracle set A. The oracle machine M operates in the same way as ordinary TMs if it is not in the query state. When it enters the query state, the oracle A takes over and performs the following tasks: it reads the query y from the query tape, cleans up the query tape, puts the head of the query tape to the original position, and puts machine M into the yes state if $y \in A$ or puts it into the no state if $y \notin A$. All of these actions made by the oracle count only one step of machine move. Let M be an oracle machine and A be a set. We write M^A to denote the computation of machine M using A as the oracle and write $L(M, A)$ to denote the set of strings accepted by M^A.

The time and space complexity of an oracle machine is defined in a natural way. Namely, for any fixed oracle A, M^A has time (space) complexity $\leq f(n)$ if for all x, $M^A(x)$ halts in $\leq f(|x|)$ moves (or, respectively, if $M^A(x)$ halts using $\leq f(|x|)$ cells, including the space of the query tape[3]). We say M has time (space) complexity $\leq f(n)$ if for all oracles A, M^A has time (space) complexity $\leq f(n)$. M is said to have a polynomial time (space) complexity if M has time (space) complexity $\leq p(n)$ for some polynomial function p. The complexity classes $P(A)$ and $PSPACE(A)$ can be defined as follows:

$$P(A) = \{L(M, A) \mid M^A \text{ has a polynomial time complexity}\},$$
$$PSPACE(A) = \{L(M, A) \mid M^A \text{ has a polynomial space complexity}\}.$$

The class $NP(A)$ may be defined to be the class of sets which are computed

[2] See Section 9 for references.

[3] The definition of space complexity of an oracle machine can vary depending upon whether we include the space of the query tape in the space measure. The different definitions may result in very different types of separation results. See, for example, Buss [1986] and Wilson [1988] for detailed discussions. In this paper, we are concerned only with polynomial space. Our definition allows the machine to query only about strings of polynomially-bounded length which is a natural constraint.

by nondeterministic oracle machines in polynomial time. Here, we avoid the formal definition of nondeterministic machines and define this class in terms of polynomial-time predicates. A predicate σ of a set variable A and a string variable x is called a polynomial-time predicate, or a P^1-predicate if there exist an oracle TM M and a polynomial p such that M has time complexity $\leq p(n)$ and for all sets A and all strings x, M^A accepts x iff $\sigma(A;x)$ is true. It is clear that $B \in P(A)$ iff there exists a P^1-predicate σ such that for all x, $x \in B \iff \sigma(A;x)$ holds. It is well known that the set $NP(A)$ can be characterized as follows: a set B is in $NP(A)$ iff there exist a P^1-predicate σ and a polynomial q such that $x \in B \iff (\exists y, |y| \leq q(|x|))\ \sigma(A;\langle x,y\rangle)$. In the rest of the paper, we will write $\exists_p y$ (or $\forall_p y$) to denote $(\exists y, |y| \leq q(|x|))$ (or, respectively, $(\forall y, |y| \leq q(|x|))$), if the polynomial bound q is clear in the context or if the exact bound is irrelevant. Using this notation, we define a $\Sigma_0^{P,1}$-predicate to be a P^1-predicate, and a $\Sigma_k^{P,1}$-predicate, for $k \geq 1$, to be a predicate having the form

$$\tau(A;x) \equiv (\exists_p y_1)(\forall_p y_2)\cdots(Q_k y_k)\ \sigma(A;\langle x, y_1, \cdots y_k\rangle),$$

where σ is a P^1-predicate and $Q_k = \exists_p$ if k is odd and $Q_k = \forall_p$ if k is even (this notation will be used through out the paper). Now the relativized polynomial-time hierarchy can be defined as follows:

$$\Sigma_k^P(A) = \{L \mid (\exists \Sigma_k^{P,1}\text{-predicate } \sigma)[x \in L \iff \sigma(A;x)]\},$$
$$\Pi_k^P(A) = \{L \mid \overline{L} \in \Sigma_k^P(A)\}, \qquad k \geq 1.$$

In addition to the polynomial-time hierarchy, we are also interested in complexity classes defined by probabilistic machines. Here we avoid the precise definition of probabilistic machines but use probabilistic quantifiers to define the related complexity classes. Let q be a polynomial and σ a predicate of a set variable and a string variable. We write $(\exists^+ y, |y| \leq q(|x|))\ \sigma(A;\langle x,y\rangle)$ to denote the predicate which states that for more than 3/4 of strings y, $|y| \leq q(|x|)$, $\sigma(A;\langle x,y\rangle)$ is true. When the polynomial q is known or is irrelevant, we use the abbreviation $\exists_p^+ y$ for $(\exists^+ y, |y| \leq q(|x|))$.

The most important probabilistic complexity classes are the class R of sets computable by polynomial time probabilistic machines with one-sided errors and the class BPP of sets computable by polynomial time probabilistic machines with bounded two-sided errors. The class BPP can be naturally generalized to the BP operator. Namely, for every complexity class $\mathcal{C}$, we may define $BP\mathcal{C}$ as follows: a set L is in $BP\mathcal{C}$ if there exists a set $B \in \mathcal{C}$ such that for all x, $x \in L \Rightarrow (\exists_p^+ y)[\langle x,y\rangle \in B]$

and $x \notin L \Rightarrow (\exists^+_p y)[\langle x, y\rangle \notin B]$. In particular, we are interested in the following relativized probabilistic polynomial hierarchy:

$$R(A) = \{L \mid (\exists P^1\text{-predicate } \sigma)[x \in L \Rightarrow (\exists^+_p y)\ \sigma(A; \langle x, y\rangle)]$$
$$\text{and } [x \notin L \Rightarrow (\forall_p y) \text{ not } \sigma(A; \langle x, y\rangle)]\}.$$
$$BP\Sigma^P_k(A) = \{L \mid (\exists \Sigma^{P,1}_k\text{-predicate } \sigma)[x \in L \Rightarrow (\exists^+_p y)\ \sigma(A; \langle x, y\rangle)]$$
$$\text{and } [x \notin L \Rightarrow (\exists^+_p y) \text{ not } \sigma(A; \langle x, y\rangle)]\}, \qquad k \geq 0.$$

We let $BPH(A) = \cup_{k=0}^{\infty} BP\Sigma^P_k(A)$. It is clear that for all oracles $P(A) \subseteq R(A) \subseteq \Sigma^P_1(A)$, and $\Sigma^P_k(A) \subseteq BP\Sigma^P_k(A) \subseteq \Pi^P_{k+1}(A)$ for all $k \geq 0$. Figure 1 shows the relations between these classes. In Section 4, we will prove some separation results to show that most of these relations are the best we know to hold for all oracles.

The above complexity classes are formed by adding an $\exists^+_p$-quantifier to the alternating $\exists_p$- and $\forall_p$-quantifiers over polynomial predicates. We may also consider the hierarchy formed by alternating the $\exists^+_p$-quantifiers with the $\exists_p$-quantifiers over polynomial predicates. The resulting hierarchy is the Arthur-Merlin hierarchy of Babai [1985] or, equivalently, the interactive proof systems of Goldwasser, Micali and Rackoff [1985]. We will follow the literature and use the term "Arthur-Merlin."

$$AM_k(A) = \{L \mid (\exists P^1\text{-predicate } \sigma)$$
$$[x \in L \Rightarrow (\exists^+_p y_1)(\exists_p y_2)\cdots(Q'_k y_k)\ \sigma(A; \langle x, y_1, y_2, \cdots, y_k\rangle) \text{ and}$$
$$[x \notin L \Rightarrow (\exists^+_p y_1)(\forall_p y_2)\cdots(Q''_k y_k)\ \sigma(A; \langle x, y_1, y_2, \cdots, y_k\rangle)\},$$
$$MA_k(A) = \{L \mid (\exists P^1\text{-predicate } \sigma)$$
$$[x \in L \Rightarrow (\exists_p y_1)(\exists^+_p y_2)\cdots(Q'_{k+1} y_k)\ \sigma(A; \langle x, y_1, y_2, \cdots, y_k\rangle) \text{ and}$$
$$[x \notin L \Rightarrow (\forall_p y_1)(\exists^+_p y_2)\cdots(Q''_{k+1} y_k)\ \sigma(A; \langle x, y_1, y_2, \cdots, y_k\rangle)\},$$

where Q'_k is $\exists^+_p$ if k is odd and is $\exists_p$ if k is even, and Q''_k is $\exists^+_p$ if k is odd and is $\forall_p$ if k is even. This hierarchy is one of the very few that were known to collapse. It is known that $AM_1(A) = BPP(A)$, $MA_1(A) = NP(A)$, $MA_2(A) \subseteq AM_2(A)$, and $AM_k(A) = MA_k(A) = AM_2(A) = BP\Sigma^P_1(A)$ for all oracles A if $k \geq 3$. For more discussions about the complexity classes definable by the $\exists^+_p$ quantifier and the AM-hierarchy, see Zachos [1986].

It is well known that the class $PSPACE(A)$ can also be defined using alternating $\exists_p$- and $\forall_p$-quantifiers. That is, a set L is in $PSPACE(A)$ iff there exist a polynomial q and a P^1-predicate σ such that for all x, $x \in L$ iff

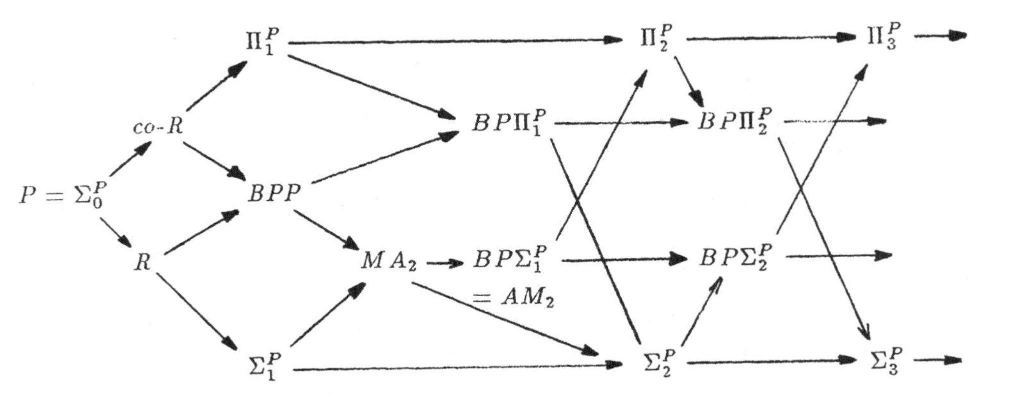

Figure 1. Polynomial and probabilistic polynomial hierarchies.

$(\exists_p y_1)(\forall_p y_2)\cdots(Q_{q(|x|)}y_{q(|x|)})\ \sigma(A; \langle x, y_1, y_2, \cdots y_{q(|x|)}\rangle)$. This suggests that if we replace the fixed integer k by a fixed polynomial bound $q(n)$ where n is the length of the input, then we obtain a generalized polynomial hierarchy which locates between the polynomial hierarchy and $PSPACE(A)$. More formally, we define, for any function $f(n)$ such that $f(n) \leq q(n)$ for some polynomial q, the following class:

$$\Sigma_{f(n)}^P(A) = \{L \mid (\exists P^1\text{-predicate } \sigma)(\forall x)[x \in L \iff (\exists_p y_1)(\forall_p y_2)\cdots(Q_{f(|x|)}y_{f(|x|)})\ \sigma(A; \langle x, y_1, \cdots y_{f(|x|)}\rangle)]\}.$$

Similar generalization can be done on the AM-hierarchy. However, the equivalence between the generalization by quantifiers and the generalization by machine models is no longer trivial. We postpone this definition to Section 7.

2. Oracle Computation and Circuits

The construction of oracles separating or collapsing hierarchies usually involves counting arguments about the computation of oracle machines. When a complex machine model, such as the nondeterministic machine or the alternating machine, is used, the computation tree is often too complicated to be comprehended. Mathematical induction has been used to simplify the arguments about the computation trees. Still, from time to time, the computation trees become so complicated that they seem beyond our understanding (cf. Baker and Selman [1979]). In addition to the mathematical induction, an important progress on simplifying the arguments about computation trees of oracle machines is to view the oracle computation tree as a circuit. This technique translates unstructured oracle computation trees into more structured, more uniform circuit computation trees, and facilitates the lower bound arguments. We illustrate this technique in this section.

A circuit is usually defined as a directed acyclic graph. For our purpose, we simply define it as a rooted tree. Each interior node of the tree is attached with a gate, and has an unlimited number of child nodes. Three types of gates will be used here: the AND gate, the OR gate and the MAJ gate (standing for "majority"). The MAJ gate outputs 1 if at least 3/4 of inputs are 1, outputs 0 if at least 3/4 of inputs are 0, and is undefined otherwise (for convenience, we say the undefined output is ?). Each leaf is attached with a constant 0, a constant 1, a variable x, or a negated variable $\bar{x}$. That is, all negation gates are moved down to the bottom by De Morgan's law. Each circuit C having only OR and AND gates has a dual circuit $\tilde{C}$ which can be defined inductively as follows: the dual of a constant or a variable is its negation, the dual of a circuit C which is an OR (or, AND) of n children C_i, $1 \leq i \leq n$, is the AND (or, respectively, OR) of $\tilde{C}_i$, $1 \leq i \leq n$.

Each circuit computes a function on its variables. In this paper, each variable is represented by v_z for some string $z \in \{0,1\}^*$. Let V be the set of variables occurred in a circuit C. Then a *restriction* ρ of C is a mapping from V to $\{0,1,*\}$. For each restriction ρ of C, $C\lceil_\rho$ denotes the circuit C' obtained from C by replacing each variable x with $\rho(x) = 0$ by 0 and each y with $\rho(y) = 1$ by 1. Assume that ρ' is a restriction of $C\lceil_\rho$. We write $C\lceil_{\rho\rho'}$ to denote $(C\lceil_\rho)\lceil_{\rho'}$. We also write $\rho\rho'$ to denote the combined restriction on C with values $\rho\rho'(x) = \rho(x)$ if $\rho(x) \neq *$ and with values $\rho\rho'(x) = \rho'(x)$ if $\rho(x) = *$. If a restriction ρ of C maps no variable to $*$, then we say ρ is an *assignment* of C. Let ρ be a restriction of C, we say that ρ *completely determines* C if $C\lceil_\rho$ computes a constant function 0 or 1. An assignment ρ of C always completely determines the circuit C. Note that we represent each variable v_z by a string $z \in \{0,1\}^*$. Therefore, for every set $A \subseteq \{0,1\}^*$, there is a natural assignment ρ_A on all variables v_z, $z \in \{0,1\}^*$: $\rho_A(v_z) = 1$ if $z \in A$ and $\rho_A(v_z) = 0$ if $z \notin A$.

Let M be a polynomial time-bounded deterministic oracle TM, and let x be an input string to M. Without knowing the answers from the oracle A, we can represent the computation of $M^A(x)$ as a tree, where each computation path corresponds to a sequence of possible answers to some queries made by M. Since the runtime of M is bounded by a polynomial p, each path consists of at most $p(|x|)$ many queries, and there are at most $2^{p(|x|)}$ many paths. Each path ends after $p(|x|)$ moves and either accepts or rejects x. For each accepting path π, let $Y_\pi = \{y | \ y$ is queried in the path π and receives the *yes* answer$\}$, and $N_\pi = \{y | \ y$ is queried in the path π and receives the *no* answer$\}$. Then, for any oracle A, $Y_\pi \subseteq A$ and $N_\pi \subseteq \overline{A}$ imply $M^A(x)$ accepts. More precisely, $M^A(x)$ accepts iff there exists an accepting path π such that $Y_\pi \subseteq A$ and $N_\pi \subseteq \overline{A}$.

Now define the circuit $C_{M,x}$ as follows. Assume that the computation tree of $M^A(x)$ has r accepting paths. Then, $C_{M,x}$ has a top OR gate with r children, each being an AND gate and corresponding to an accepting path of the computation tree of $M^A(x)$. For each path π, the corresponding AND gate has the following children: $\{v_y | \ y \in Y_\pi\} \cup \{\overline{v}_y | \ y \in N_\pi\}$. Then, circuit $C_{M,x}$ satisfies the following property: $C_{M,x}\lceil_{\rho_A}$ outputs 1 iff $M^A(x)$ accepts. In summary, we have

Lemma 2.1. Let M be an oracle TM with runtime $\leq p(n)$. Then, for each x, there is a depth-2 circuit $C = C_{M,x}$ satisfying the following properties:

(a) C is an OR of ANDs,

(b) the top fanin of C is $\leq 2^{p(|x|)}$ and the bottom fanin of C is $\leq p(|x|)$, and

(c) for any set A, $C\lceil_{\rho_A} = 1$ iff $M^A(x)$ accepts.

Remark 2.2. The above also holds if we require that the circuit C be an AND of ORs. This can be seen by considering the machine M' which computes the complement of $L(M, A)$, and noting that the dual circuit of the circuit $C_{M',x}$ satisfies properties (b) and (c).

From the above basic relation, we can derive relations regarding other complexity classes. Let τ be a $\Sigma_k^{P,1}$-predicate; that is, $\tau(A; x) \equiv (\exists_p y_1) \cdots (Q_k y_k)$ $\sigma(A; \langle x, y_1, \cdots, y_k\rangle)$ for some P^1-predicate σ. Let q be a polynomial bounding the length of y_i, $1 \leq i \leq k$, as well as the runtime of the oracle machine for σ. When translating this predicate into a circuit, it is natural to identify an AND gate with a $\forall_p$-quantifier and an OR gate with an $\exists_p$-quantifier. We call a depth-$(k+1)$ circuit a Σ_k-*circuit* if it has alternating OR and AND gates, starting with a top OR gate. A Σ_k-circuit is called a $\Sigma_k(m)$-*circuit* if its fanins are $\leq 2^m$ and the bottom fanins are $\leq m$. (Note that a Σ_k-circuit has depth $k+1$ rather than k, because a $\Sigma_k^{P,1}$-predicate corresponds to a depth-$(k+1)$ circuit.) A Π_k-*circuit* is the dual circuit of a Σ_k-circuit and a $\Pi_k(m)$-*circuit* is the dual circuit of a $\Sigma_k(m)$-circuit. Then we have the following relations.

Lemma 2.3. Let $k \geq 1$. For every $\Sigma_k^{P,1}$-predicate τ there is a polynomial q such that for every x, there exists a $\Sigma_k(q(|x|))$-circuit $C_{\tau,x}$, having the property that for any set A, $C_{\tau,x}\lceil_{\rho_A}= 1$ iff $\tau(A; x)$ is true. The similar relation holds between $\Pi_k^{P,1}$-predicates and Π_k-circuits.

Note that the depth of the above circuit is $k + 1$ rather than $k + 2$ because the gate corresponding to the last quantifier can always be combined with the top gate of the circuit $C = C_{\sigma,\langle x,y_1,\cdots y_k\rangle}$: depending upon whether the last quantifier is an $\exists_\mathrm{p}$ (when k is odd) or a $\forall_\mathrm{p}$ (when k is even), the circuit C may be made to be an OR of ANDs (as in Lemma 2.1) or an AND of ORs (as in Remark 2.2), respectively. Also note that in Lemma 2.3, the circuit $C_{\tau,x}$ exists for every x, therefore we may let the parameter k be a function of x. In other words, the above lemma generalizes immediately to the generalized polynomial hierarchy $\Sigma_{f(n)}^P(A)$.

We can also extend the above lemma to the BP-hierarchy. For each $\exists_\mathrm{p}^+$-quantifier, the natural corresponding gate is a MAJ gate. Hence, we have the following relation between circuits with MAJ gates and probabilistic complexity classes. A circuit C is a *$BP\Sigma_k(m)$-circuit* if it is an MAJ gate having $\leq 2^m$ many $\Sigma_k(m)$-circuits as children.

Lemma 2.4. For every $BP\Sigma_k^{P,1}$-predicate τ there is a polynomial q such that for every x, there exists a $BP\Sigma_k(q(|x|))$-circuit $C_{\tau,x}$, having the property that for any set A, $C_{\tau,x}\lceil_{\rho_A}= 1$ if $\tau(A; x)$ is true and $C_{\tau,x}\lceil_{\rho_A}= 0$ if $\tau(A; x)$ is false.

3. Diagonalization

Let $\mathcal{C}_1$ and $\mathcal{C}_2$ be two complexity classes. A standard proof technique of constructing oracles A to separate $\mathcal{C}_1$ from $\mathcal{C}_2$ (so that $\mathcal{C}_1(A) \not\subseteq \mathcal{C}_2(A)$) is the technique of diagonalization. In a proof by diagonalization, we consider a set $L_A \in \mathcal{C}_1(A)$ against all sets in $\mathcal{C}_2(A)$ and, at each stage, extend set A on a finite number of strings so that L_A is not equal to the specific set in $\mathcal{C}_2(A)$ under consideration in this stage. Therefore, after we considered all sets in $\mathcal{C}_2(A)$, it is established that L_A is not in $\mathcal{C}_2(A)$. We illustrate this technique in this section.

We begin by looking at a simple example: separating the class $\mathcal{C}_1 = NP = \Sigma_1^P$ from the class $\mathcal{C}_2 = co\text{-}NP = \Pi_1^P$. First, we need to enumerate all sets in $co\text{-}NP(A)$. We do it by considering an enumeration of all $\Sigma_1^{P,1}$-predicates $\{\sigma_i\}$. We assume that the ith $\Sigma_1^{P,1}$-predicate $\sigma_i(A;x)$ is of the form $(\exists y, |y| \le q(|x|))\tau_i(A;\langle x,y\rangle)$ such that τ_i is a P^1-predicate whose runtime is bounded by the ith polynomial $p_i(|x|)$ and that $q(n) \le p_i(n)$. Let

$$L_A = \{0^n \mid (\exists y, |y| = n)\ y \in A\}.$$

Then, for all A, $L_A \in NP(A)$. We will construct set A by stages such that in stage n, the following requirement R_n is satisfied:

R_n: $(\exists x_n)\ [x_n \in L_A \iff \sigma_n(A;x_n)]$.

Observe that for every set $B \in \Pi_1^P(A)$, there must be a $\Sigma_1^{P,1}$-predicate σ_i such that for all x, $x \in B$ iff *not* $\sigma_i(A;x)$. Now, if requirement R_i is satisfied then $x_i \in L_A \iff \sigma_i(A;x_i) \iff x_i \notin B$, and hence $L_A \ne B$. Thus, if all requirements R_n are satisfied, then $L_A \notin \Pi_1^P(A)$.

We now describe the construction of set A. Set A will be constructed by stages. In each stage, we will reserve some strings for set A and some strings for set $\overline{A}$. We let $A(n)$ and $A'(n)$ be sets containing those strings reserved for A and $\overline{A}$, respectively, up to stage n. We begin with sets $A(0) = A'(0) = \emptyset$. In stage n, we try to satisfy requirement R_n. Assume that before stage n we have defined $t(n-1)$ such that $A(n-1) \cup A'(n-1) \subseteq \Sigma^{\le t(n-1)}$. We let m be the least integer greater than $t(n-1)$ such that $2^m > p_n(m)$, and let $x_n = 0^m$. Now we consider the computation tree generated by $\sigma_n(A;x_n)$, with the following modification: if the tree contains a query "$y \in ?A$" with $|y| < m$, then prune the *no* child if $y \in A(n)$ and prune the *yes* child if $y \notin A(n)$. Thus we obtain a computation tree of queries "$y \in ?A$" only if $|y| \ge m$. Consider two cases:

Case 1. There exists an accepting path π in the tree. Then, this accepting path π is of length at most $p_n(m)$ and hence asks at most $p_n(m)$ many queries about strings y of length m. Since $2^m > p_n(m)$, there must be at least one string z of length m not being queried in the path π. We fix such a string z_0. Also, let $B_0 = \{y \mid y$ is queried and answered *no* in path $\pi\}$, and $B_1 = \{y \mid y$ is queried and answered *yes* in path $\pi\}$. We define $A(n) = A(n-1) \cup B_1 \cup \{z_0\}$ and $A'(n) = A'(n-1) \cup B_0$. Note that requirement R_n is satisfied by set $A(n)$ and string x_n: $x_n \in L_{A(n)}$ because $|z_0| = m$ and $z_0 \in A(n)$, and $\sigma_n(A(n);x_n)$ because $B_1 \subseteq A(n)$ and $B_0 \cap A(n) = \emptyset$ imply that the computation of $\sigma_n(A(n);x_n)$ follows the path π and accepts.

Case 2. All paths in the tree reject. Then, consider the path π all of whose queries "$y \in ?A$" receive the answer *no*. Let $B_0 = \{y \mid y$ is queried in the path $\pi\}$. We let $A(n) = A(n-1)$ and $A'(n) = A'(n-1) \cup B_0$. Note that requirement R_n is satisfied by set $A(n)$ and string x_n: $x_n \notin L_{A(n)}$ because $A(n) \cap \{0,1\}^m = \emptyset$, and *not* $\sigma_n(A(n); x_n)$ because $B_0 \cap A(n) = \emptyset$ implies that the computation of $\sigma_n(A(n); x_n)$ follows the path π and rejects.

Finally we complete stage n by letting $t(n) = p_n(m)$. Note that for all y in $A(n) \cup A'(n)$, $|y| \leq t(n)$ (assuming $p_n(m) > m$).

We define set A to be $\cup_{n=0}^{\infty} A(n)$. Note that the set A has the property that $A^{\leq t(n)} = A(n)$. Thus, the requirement R_n is satisfied by set A and string x_n because both computations of $x_n \in L_A$ and $\sigma_n(A; x_n)$ involve only with strings of length $\leq t(n)$ and so $A^{\leq t(n)} = A(n)$ implies that requirement R_n is satisfied by A and x_n.

In the above we have described in detail the construction of a set A such that $NP(A) \neq co\text{-}NP(A)$. Now we re-examine the construction and make the following observations about its general properties.

(1) We need an enumeration of sets in C_2. This is usually simple. For the classes we defined in Section 1, all of them have simple representations by a number of polynomial length-bounded quantifiers followed by P^1-predicates. Since the class of all P^1-predicates have a simple enumeration, we can enumerate these classes accordingly. Namely, we may assume, for any sequence of polynomial length-bounded quantifiers $Q'_1, \cdots, Q'_k$, an enumeration $\{\sigma_i\}$ of predicates of the form $(Q'_1 y_1) \cdots (Q'_k y_k)\ \tau(A; \langle x, y_1, \cdots, y_k \rangle)$ such that both the runtime of the P^1-predicate τ and the length of y_j's are bounded by the ith polynomial p_i.

(2) The requirement $C_1(A) \not\subseteq C_2(A)$ is divided into an infinite number of requirements:

$$R_n\text{: } (\exists x_n)[x_n \in L_A \iff \textit{not } \sigma_n(A; x_n)],$$

where L_A is a fixed set in $C_1(A)$ and σ_n is the nth predicate in our enumeration of sets in $C_2(A)$.

(3) Assume that $A(n) = A^{\leq t(n)}$ and let $D(n) = \{y \mid t(n-1) < |y| \leq t(n)\}$. We call set $D(n)$ the *diagonalization region* for stage n. Then, sets $D(n)$ and $A(n)$ satisfy

(a) $x_n \in L_A \iff x_n \in L_{A(n)} \iff x_n \in L_{A \cap D(n)}$, and

(b) $\sigma_n(A; x_n) \iff \sigma_n(A(n); x_n) \iff \sigma_n(A(n-1) \cup (A \cap D(n)); x_n)$.

Therefore, in stage n we essentially only need to satisfy the following simpler requirement R'_n, instead of R_n.

$$R'_n\text{: } (\exists x_n)(\exists B \subseteq D(n))[x_n \in L_B \iff \textit{not } \sigma_n(A(n-1) \cup B; x_n)].$$

(4) Inside the diagonalization region $D(n)$, we need to show that requirement R'_n can be satisfied, usually by a counting argument. We observe that the counting argument used in the above example is essentially equivalent to a lower bound argument for $\Sigma_1(m)$-circuits versus $\Pi_1(p_n(m))$-circuits. More precisely, in stage n, the computation tree of $\sigma_n(A; x_n)$, after pruning to remove queries about strings of length less than m, can be translated to a $\Sigma_1(p_n(m))$-circuit C (Lemma 2.3). Let $\tilde{C}$ be its dual circuit. Then $\tilde{C}$ is a $\Pi_1(p_n(m))$-circuit. Also, the question of whether $x \in L_A$ is equivalent to the predicate $(\exists z, |z| = m)[z \in A]$ and hence can be expressed by a simple $\Sigma_1(m)$-circuit C_0: C_0 is the OR of 2^m variables v_z, $|z| = m$. Now observe that the counting argument in the above example essentially establishes that there is an assignment ρ on variables such that $C_0\lceil_\rho \neq \tilde{C}\lceil_\rho$.

In general, we can see that the requirement R'_n can be reduced to the requirement R''_n about circuits:

R''_n: $(\exists x_n)(\exists B \subseteq D(n))[x_n \in L_B \iff C\lceil_{\rho_B} = 0]$,

where C is the circuit corresponding to predicate $\sigma_n(A; x_n)$, with each variable v_y having $|y| \leq t(n-1)$ replaced by value $\chi_{A(n-1)}(y)$. Depending upon which class $\mathcal{C}_1$ is, the predicate $x_n \in L_B$ may also be represented by a circuit C_0. In this case, requirement R''_n states that $(\exists x_n)(\exists B \subset D(n))[C_0\lceil_{\rho_B} \neq C\lceil_{\rho_B}]$.

The above discussion shows that whenever possible, the diagonalization process is reduced to a lower bound problem about circuits. In the next section, we will see more examples using the above setting of diagonalization.

4. Separation Results

In this section, we use the general setting of diagonalization discussed in Section 3 to separate some complexity hierarchies defined in Section 1 by oracles.

4.1. Separating PSPACE from PH

First, we consider the case when $\mathcal{C}_1 = PSPACE$ and $\mathcal{C}_2 = \Sigma_k^P$, for arbitrary $k \geq 1$. We set up the following items:

(a) $L_A = \{0^n \mid \|A^{=n}\| \text{ is odd}\} \in PSPACE(A)$.

(b) For each n, let m be a sufficiently large integer greater than $t(n-1)$ (exact bound for m to be determined later). Let $x_n = 0^m$ and $t(n) = p_n(m)$, and recall that $D(n) = \{y \mid t(n-1) < |y| \leq t(n)\}$.

(c) For each n, let σ_n be the nth $\Sigma_k^{P,1}$-predicate and let C_n be the $\Sigma_k(p_n(m))$-circuit such that for all sets $B \subseteq D(n)$, $C_n\lceil_{\rho_B} = 1$ iff $\sigma_n(A(n-1) \cup B; x_n)$. (That is, C_n is obtained from predicate $\sigma_n(A; x_n)$ by Lemma 2.3 with the modification that each variable v_y is assigned value $\chi_{A(n-1)}(y)$ if $|y| < m$.)

Now the separation problem for *PSPACE* versus Σ_k^P is reduced to the following requirement:

R_n'': $(\exists B \subseteq D(n))$ $[\|B^{=m}\|$ is odd $\iff C_n\lceil_{\rho_B} = 0]$.

Or, equivalently, it is reduced to the following lower bound result on parity circuits. For each $k \geq 1$, let $s_k(n)$ be the minimum size r of a depth-k circuit[4] which has size r and bottom fanin $\leq \log r$ and computes the (odd) parity of n variables. Requirement R_n'' states that the function $s_k(n)$ grows faster than the functions $2^{(\log n)^k}$ for all k. In the following we show an even stronger result: $s_k(n)$ is greater than an exponential function on n. This is the main breakthrough in the theory of relativization.

Theorem 4.1 [Yao, 1985; Hastad, 1987]. For sufficiently large n, $s_k(n) \geq 2^{(1/10)n^{1/(k-1)}}$

Since the proof of this theorem is quite involved, we will only include a sketch of the main ideas here. The interested reader is referred to Hastad [1987] for the complete proof.

Proof of Theorem 4.1. The proof is done by induction on k. A stronger statement is easier to use in the induction proof. More precisely, we are going to show the following stronger form of the theorem:

Induction statement. Let $k \geq 2$ and $\delta = 1/10$. Let C_n be a depth-k circuit having $\leq 2^{\delta n^{1/(k-1)}}$ gates not at the bottom level and having bottom fanin $\leq \delta n^{1/(k-1)}$. Then, for sufficiently large n, C_n does not compute the parity of n variables.

The base case of $k = 2$ is very easy to see as no depth-2 circuit having bottom fanin $\leq n-1$ computes the parity of n variables.

For the inductive step, assume that $k > 2$. By way of contradiction, we assume that there exists a circuit C_n of the above form that computes the parity of

[4] In this and the next subsections, all circuits are circuits that have only OR and AND gates.

n variables. To apply the inductive hypothesis, we need to show that there exists a restriction ρ such that it leaves m variables v unassigned (i.e., $\rho(v) = *$) but makes circuit $C_n\lceil_\rho$ to be equivalent to a depth-$(k-1)$ circuit having $\leq 2^{\delta m^{1/(k-1)}}$ gates not at the bottom level and having bottom fanin $\leq \delta m^{1/(k-1)}$. Then, the assumption that C_n computes the parity of n variables implies that $C_n\lceil_\rho$ (or its dual circuit) computes the parity of m variables, which leads to a contradiction to the inductive hypothesis.

To this end, we consider the following probability space R_p of restrictions on the n variables: a random restriction ρ from R_p satisfies that $\Pr[\rho(v) = *] = p$ and $\Pr[\rho(v) = 0] = \Pr[\rho(v) = 1] = (1-p)/2$, where p is a parameter, $0 \leq p \leq 1$. Then, we consider the circuit $C_n\lceil_\rho$ resulted from applying a random restriction ρ from R_p to C_n.

If we take p, for instance, to be $n^{-1/(k-1)}$, then the expected number of variables assigned with $*$ by ρ is $n^{(k-2)/(k-1)}$. Therefore, for probability $\geq 1/3$, a random restriction ρ will leave more than $n^{(k-2)/(k-1)}$ variables unassigned. To apply the inductive hypothesis, we need only to show that the probability is greater than $2/3$ that the circuit $C_n\lceil_\rho$ is equivalent to a depth-$(k-1)$ circuit having $\leq 2^{\delta r}$ many gates not at the bottom level and having bottom fanin $\leq \delta r$, where $r = (n^{(k-2)/(k-1)})^{1/(k-2)} = n^{1/(k-1)}$, and ρ is a random restriction from R_p with $p = n^{-1/(k-1)}$. This is the consequence of the following Switching lemma:

Lemma 4.2 (Switching Lemma I). Let G be a depth-2 circuit such that G is the AND of ORs with bottom fanin $\leq t$ and ρ a random restriction from R_p. Then, the probability that $G\lceil_\rho$ is not equivalent to an OR of ANDs with bottom fanin $\leq s$ is bounded by α^s, where α satisfies $\alpha \leq 5pt$. The above also holds if G is an OR of ANDs to be converted to a circuit of AND of ORs.

The proof of the Switching Lemma is too complicated to be included here. We omit it and continue the induction proof for Theorem 4.1. Without loss of generality, assume that k is odd and so the bottom gate of C_n is an OR gate. Let $s = t = \delta n^{1/(k-1)}$. Note that C_n has $\leq 2^s$ many depth-2 subcircuits. From the Switching Lemma, the probability that every bottom depth-2 subcircuit G of $C_n\lceil_\rho$ (which is an AND of ORs) is equivalent to a depth-2 circuit of OR of ANDs, with bottom fanins $\leq s$, is

$$\geq (1-\alpha^s)^{2^s} \geq 1 - (2\alpha)^s.$$

Now, $\alpha < 5pt = 1/2$ and so $\lim_{n\to\infty}(2\alpha)^s = 0$. In particular, there exists an integer

n_k such that if $n > n_k$ then $(2\alpha)^s < 1/3$. (This integer n_k depends only on k.) Thus, for sufficiently large n, the probability is greater than $1 - (2\alpha)^s \geq 2/3$ that $C_n\lceil_\rho$ is equivalent to a depth-$(k-1)$ circuit having $\leq 2^{\delta n^{1/(k-1)}}$ gates not at the bottom level and having bottom fanin $\leq \delta n^{1/(k-1)}$. This completes the proof of Theorem 4.1. □

Remark 4.3. The above gave the lower bound $s_k(n) \geq 2^{(1/10)n^{1/(k-1)}}$ on the size r of Σ_k-circuits C for parity if C must have bottom fanin $\leq \log r$. We may treat a Σ_k-circuit as a Σ_{k+1}-circuit with bottom fanin 1, and so we obtain the lower bound $s'_k(n) \geq 2^{(1/10)n^{1/k}}$ on the size of any Σ_k-circuit for parity. Hastad [1987] has pointed out that we may also first apply the Switching Lemma to shrink the bottom fanin to $\log r$ and so obtain a better bound $s'_k(n) \geq 2^{\epsilon n^{1/(k-1)}}$, where $\epsilon = 10^{-k/(k-1)}$.

From Theorem 4.1, it is clear that if we choose integer m to be large enough such that Theorem 4.1 holds for $s_k(2^m)$ and that $(1/10)2^{m/(k-1)} > k \cdot p_n(m)$ (note that $2^{k \cdot p_n(m)}$ bounds the size of a $\Sigma_k(p_n(m))$-circuit), then requirement R''_n can be satisfied. By dovetailing the enumeration of $\Sigma_k^{P,1}$-predicates for all $k \geq 1$, we obtain the following theorem.

Theorem 4.4. There exists an oracle A such that for every $k \geq 1$, $PSPACE(A) \neq \Sigma_k^P(A)$.

Let $\oplus P(A)$ (read "parity-$P(A)$) be the complexity class defined as follows: a set L is in $\oplus P(A)$ iff there exists a P^1-predicate σ such that for all x, $x \in L$ iff the number of y, $|y| \leq q(|x|)$, such that $\sigma(A; \langle x, y\rangle)$ holds is odd. It is easy to see that $\oplus P(A) \subseteq PSPACE(A)$ for all A, but it is not known whether $NP \subseteq \oplus P$ or $\oplus P \subseteq \Sigma_k^P$ for any $k \geq 1$. It is easy to see that the set L_A in the above proof is in $\oplus P(A)$. That is, we have actually established a stronger separation result.

Corollary 4.5. There exists a set A such that $\oplus P(A) \not\subseteq \Sigma_k^P(A)$ for all $k \geq 0$.

4.2. Separating PH

In this subsection, we show that there is an oracle A such that $PH(A)$ is an infinite hierarchy. It is sufficient to find, for each $k > 0$, a set $L_A \in \Sigma_k^P(A)$ such that $L_A \notin \Pi_k^P(A)$. Then, by dovetailing the diagonalization for each $k > 0$, the oracle A can be constructed so that $\Sigma_k^P(A) \neq \Pi_k^P(A)$ for all k.

Let $k > 0$ be fixed. Define

$$L_A = \{0^{(k+1)n} \mid (\exists y_1, |y_1| = n)(\forall y_2, |y_2| = n) \cdots (Q_k y_k, |y_k| = n)\ 0^n y_1 y_2 \cdots y_k \in A\}.$$

Then, clearly, $L_A \in \Sigma_k^P(A)$. The setup for m, $t(n)$ and $D(n)$ is similar to that in Section 4.1. (In particular, m must be *sufficiently* large.) From the general setting discussed in Section 3, we only need to satisfy the requirements

$$R_n'': (\exists x_n = 0^m)(\exists B \subseteq D(n))[0^m \in L_B \iff C\lceil_{\rho_B} = 0],$$

where C is a $\Pi_k(p_n(|m|))$-circuit corresponding to the nth $\Pi_k^{P,1}$-predicate $\sigma_n(A; 0^m)$, with all variables v_y having $|y| \leq t(n-1)$ replaced by the constant $\chi_{A(n-1)}(y)$.

We note that the predicate $0^m \in L_B$ is a $\Sigma_k^{P,1}$-predicate. That is, there is a depth-k circuit C_0 having the following properties:

(a) C_0 has alternating OR and AND gates, starting with a top OR gate,

(b) all the fanins of C_0 are exactly $2^{m/(k+1)}$,

(c) each leaf of C_0 has a unique positive variable v_z with $|z| = m$, and

(d) for all $B \subseteq D(n)$, $C_0\lceil_{\rho_B} = 1 \iff 0^m \in L_B$.

So, the requirement R_n'' is reduced to the following problem on circuits: there exists a set $B \subseteq D(n)$ such that $C_0\lceil_{\rho_B} \neq C\lceil_{\rho_B}$. We restate it in the lower bound form. Let D be a depth-k circuit having the following properties: (a) it has alternating OR and AND gates, with a top OR gate, (b) all its fanins are exactly n, except the bottom fanins which are exact $\sqrt{n}$, and (c) each leaf of D has a unique positive variable. We let the function f_k^n be the function computed by D. (Note that C_0 contains a subcircuit computing a function $f_k^{2^{m/(k+1)}}$.) Let $s_k(n)$ be the minimum r of a Π_k-circuit computing a function f_k^n that has size $\leq r$ and bottom fanin $\leq \log r$. The new requirement R_n'' is satisfied by the following lower bound on $s_k(n)$.

Theorem 4.6. For sufficiently large n, $s_k(n) \geq 2^{(1/12)n^{1/3}}$

Sketch of proof. Again, we prove it by induction. The induction is made easier on the following stronger form.

Induction statement. Let $k \geq 1$ and $\delta = 1/12$. Let C_n be a Π_k-circuit having $\leq 2^{\delta n^{1/3}}$ gates not at the bottom level and having bottom fanin $\leq \delta n^{1/3}$. Then, for sufficiently large n, C_n does not compute the function f_k^n.

The base step of the induction involves two circuits: C_n is an AND of ORs with small bottom fanins ($\leq \delta n^{1/3}$), and C_0 (which computes the function f_1^n) is an OR of $n^{1/2}$ variables. We need to show that C_n does not compute the same function as C_0. But this is exactly what we proved in the example of Section 3.

For the inductive step, consider $k > 1$. We define two new probability spaces of restrictions: $R_{q,\mathcal{B}}^+$ and $R_{q,\mathcal{B}}^-$, where $\mathcal{B} = \{B_j\}_{j=1}^r$ is a partition of variables and q a value between 0 and 1. To define a random restriction ρ in $R^+_{q,\mathcal{B}}$, first, for each B_j, $1 \leq j \leq r$, let $s_j = *$ with probability q and $s_j = 0$ with probability $1-q$; [5] and then, independently, for each variable $x \in B_j$, let $\rho(x) = s_j$ with probability q and $\rho(x) = 1$ with probability $1-q$. Next, define, for each $\rho \in R^+_{q,\mathcal{B}}$, a restriction $g(\rho)$: for all B_j with $s_j = *$, let V_j be the set of all variables in B_j which are given value $*$ by ρ; $g(\rho)$ selects one variable y in V_j and gives value $*$ to y and value 1 to all others in V_j. The probability space $R^-_{q,\mathcal{B}}$ and $g(\rho)$ are defined by interchanging the roles played by 0 and 1.

Now let $\mathcal{B} = \{B_j\}$ be the partition of variables in C_0 such that each B_j is the set of all variables leading to a bottom gate in C_0. Let $q = n^{-1/3}$. If k is even, then apply a random restriction ρ from $R^+_{q,\mathcal{B}}$; otherwise, apply a random restriction ρ from $R^-_{q,\mathcal{B}}$.

In order to apply the inductive hypothesis, we need to verify that (a) with a high probability, $C\lceil_{\rho g(\rho)}$ is equivalent to a Π_{k-1}-circuit having $\leq 2^{\delta n^{1/3}}$ gates not at the bottom level and having bottom fanin $\leq \delta n^{1/3}$, and (b) with a high probability, $C_0\lceil_{\rho g(\rho)}$ contains a subcircuit computing the function f^n_{k-1}. We give sketches of these facts.

To prove part (a), we need a new version of the Switching Lemma.

Lemma 4.7 (Switching Lemma II). Let G be an AND of ORs with bottom fanin $\leq t$, and $\mathcal{B} = \{B_j\}$ be a partition of variables in G. Then, for a random restriction ρ from $R^+_{q,\mathcal{B}}$, the probability that $G\lceil_{\rho g(\rho)}$ is not equivalent to a circuit of OR of ANDs with bottom fanin $\leq s$ is bounded by α^s, where $\alpha < 6qt$. The above also holds with $R^+_{q,\mathcal{B}}$ replaced by $R^-_{q,\mathcal{B}}$, or with G being an OR of ANDs to be converted to a circuit of an AND of ORs.

We again omit the proof of the second Switching Lemma; the interested reader is referred to Hastad [1987] for details.

Without loss of generality, assume that k is even. From Lemma 4.7, if we choose $s = t = \delta n^{1/3}$, then we have $\alpha < 6qt = 1/2$, and so $\lim_{n\to\infty}(2\alpha)^s = 0$. Therefore, the probability that every bottom depth-2 subcircuit $G\lceil_{\rho g(\rho)}$ (which is an AND of ORs) of $C\lceil_{\rho g(\rho)}$ is equivalent to a depth-2 circuit of OR of ANDs is

$$\geq (1-\alpha^s)^{2^s} \geq 1-(2\alpha)^s \geq 2/3,$$

[5] Do not confuse the s_j here (which is a value in $\{0, 1, *\}$) with the size function $s_k(n)$.

for sufficiently large n. The above proved part (a). The following lemma proves part (b).

Lemma 4.8. For sufficiently large n, the probability that $C_0\lceil_{\rho g(\rho)}$ contains a subcircuit for f_{k-1}^n is greater than 2/3.

Sketch of proof. We want to show that $q_1 q_2 \geq 2/3$, where q_1 is the probability that all bottom AND gates $H_j\lceil_{\rho g(\rho)}$ (corresponding to block B_j) of $C_0\lceil_{\rho g(\rho)}$ takes the value s_j, and q_2 is the probability that all OR gates at level 2 from bottom of $C_0\lceil_{\rho g(\rho)}$ have $\geq n^{1/2}$ many child nodes $H_j\lceil_{\rho g(\rho)}$ of AND gates having value $s_j = *$.

To estimate probability q_1, we note that for a fixed j,

$$\begin{aligned}\Pr[H_j\lceil_{\rho g(\rho)} \text{ has value } \neq s_j] &= \Pr[\text{all inputs to } H_j\lceil_{\rho g(\rho)} \text{ are } 1]\\ &= (1-q)^{n^{1/2}} = (1-n^{-1/3})^{n^{1/2}} < e^{-n^{1/6}},\end{aligned}$$

for sufficiently large n. Since C_0 has n^{k-1} many bottom gates H_j, we obtain $q_1 \geq (1-e^{-n^{1/6}})^{n^{k-1}} \geq 5/6$, for sufficiently large n.

To estimate probability q_2, let G be a bottom depth-2 subcircuit of C_0, and let r_ℓ be the probability that $G\lceil_{\rho g(\rho)}$ has exactly ℓ many child nodes $H_j\lceil_{\rho g(\rho)}$ having values $s_j = *$. Then,

$$r_\ell = \binom{n}{\ell} q^\ell (1-q)^{n-\ell}.$$

Note that for sufficiently large n, if $\ell \leq 2\sqrt{n}$ then $r_\ell \leq 1/2$ and $r_\ell \geq 2r_{\ell-1}$. So,

$$\sum_{\ell=0}^{\sqrt{n}} r_\ell \leq r_{\sqrt{n}} \cdot \sum_{\ell=0}^{\sqrt{n}} 2^{-\ell} \leq 2 \cdot r_{\sqrt{n}} \leq 2 \cdot 2^{-\sqrt{n}} \cdot r_{2\sqrt{n}} \leq 2^{-\sqrt{n}}.$$

Thus, $q_2 \geq (1-2^{-\sqrt{n}})^{n^{k-2}} > 5/6$, for sufficiently large n. This shows that $q_1 \cdot q_2 \geq 2/3$ and completes the proof for Lemma 4.8 and hence Theorem 4.6. □

Theorem 4.6 implies that requirement R_n'' can be satisfied if we choose integer m to be so large that Theorem 4.6 holds for $s_k(2^{m/(k+1)})$ and that $(1/12)2^{m/3(k+1)} > k \cdot p_n(m)$.

Theorem 4.9. There exists an oracle A such that for all $k > 0$, $\Sigma_k^P(A) \neq \Pi_k^P(A)$.

4.3. Separating BPH from PH

We now consider the hierarchy $BPH(A) = \cup_{k=0}^{\infty} BP\Sigma_k^P(A)$. Since $\Sigma_k^P(A) \subseteq BP\Sigma_k^P(A) \subseteq \Pi_{k+1}^P(A)$, it follows immediately that $BP\Sigma_{k+1}^P(A) = BP\Sigma_k^P(A)$ implies that the polynomial hierarchy $PH(A)$ collapses to $BP\Sigma_k^P(A)$. Therefore, there exists an oracle A such that $BPH(A)$ is infinite. What we are going to show in this section is instead that there exists an oracle A such that $BP\Sigma_k^P(A) \not\subseteq \Sigma_{k+1}^P(A)$; or, in other words, the two hierarchies $PH(A)$ and $BPH(A)$ are completely separated.

We let $L_A = \{0^{(k+2)n} \mid (\exists^+ y_0, |y_0| = n)(\exists y_1, |y_1| = n)(\forall y_2, |y_2| = n) \cdots (Q_k y_k, |y_k| = n) 0^n y_0 y_1 \cdots y_k \in A\}$. Then, obviously, $L_A \in BP\Sigma_k^P(A)$. Also, for stage n, we set up $x_n = 0^m$ for appropriate value m and let C be the $\Sigma_{k+1}(p_n(m))$-circuit corresponding to the nth $\Sigma_{k+1}^{P,1}$-predicate $\sigma_n(A; 0^m)$ with variables v_y assigned value $\chi_{A(n-1)}(y)$ if $|y| < m$. To satisfy the requirement

$$R_n'': (\exists x_n = 0^m)(\exists B \subseteq D(n))[0^m \in L_A \iff C\lceil_{\rho_B} = 0],$$

we only need a lower bound result like that in Section 4.2. More precisely, let C_i, $1 \le i \le n$, be circuits computing functions f_k^n, and let C_0 be a depth-$(k+1)$ circuit with a top MAJ gate and having $C_1, \cdots, C_n$ as its n children. Then, we need to show that the minimum size r of a Σ_{k+1}-circuit that has bottom fanin $\le \log r$ and is equivalent to C_0 is at least $2^{(1/12)n^{1/3}}$. If we treat the top MAJ gate of C_0 as an AND gate, then the proof for this lower bound result is almost identical to the proof given in Section 4.2. The only difference is that the base step of the induction proof is no longer simple. In fact, it needs a more complicated counting argument. We state it as a separate lemma.

Lemma 4.10. Let C_0 be a depth-2 circuit having a top MAJ gate with n children, each being an OR of $n^{1/2}$ many variables, and let C be a $\Sigma_3(m)$-circuit, where $m = (1/12)n^{1/3}$. Then, for sufficiently large n, there exists an assignment ρ such that $C_0\lceil_\rho \ne ?$ and $C_0\lceil_\rho \ne C\lceil_\rho$.

Proof. Note that we *cannot* apply a random restriction ρ from $R_{q,B}^-$ to the circuits and use the second Switching Lemma to simplify the circuits, because such a random restriction ρ would oversimplify circuit C into one that computes a constant 1 because the majority (more than 3/4 many) of the OR gates in $C\lceil_\rho$ would have value $s_j = 1$. Instead of using a random restriction to simplify circuits, we make a direct ad hoc counting argument to show that circuit C cannot simulate circuit C_0. This counting argument was first used by Baker and Selman [1979] to construct an oracle A such that $\Sigma_2^P(A) \ne \Pi_2^P(A)$. We give a sketch here.

Assume, by way of contradiction, that for all assignments ρ, $C_0\lceil_\rho \ne ?$ implies $C_0\lceil_\rho = C\lceil_\rho$. Let V be the set of variables in C_0. Without loss of generality, assume that all variables in C are in V. Let

$$K = \{\rho | \ (\forall \text{ OR gate } H \text{ of } C_0)(\exists \text{ unique } z \text{ in } H) \ \rho(z) = 1\}.$$

Then, $||K|| = n^{n/2} = 2^{(n \log n)/2}$. For each $\rho \in K$, $C_0\lceil_\rho = 1$, and hence one of AND gate D of C has $D\lceil_\rho = 1$. Note that there are $\le 2^m$ AND gates in C. So, there exists an AND gates D of C such that $D\lceil_\rho = 1$ for at least $2^{(n \log n)/2 - m}$ many ρ in K. Fix this AND gate D. Let $K_0 = \{\rho \in K | \ D\lceil_\rho = 1\}$, and $Q_0 = \emptyset$. We note that K_0 and Q_0 satisfy the following properties:

(i) $||K_\mu|| \ge 2^{(n/2 - \mu/3)\log n - c\mu - m}$, where $c = \log 12$,

(ii) $||Q_\mu|| = \mu$,

(iii) $\rho \in K_\mu \Rightarrow [\rho \in K_0 \text{ and } \rho(w) = 1 \text{ for all } w \in Q_\mu]$.

We claim that we can find sets K_μ and Q_μ satisfying the above properties for $\mu = 0, \cdots, n/4$. Then, we have $||K_{n/4}|| \ge 2^{(n/2 - n/12)\log n - cn/4 - m} = 2^{(5/12)n \log n - cn/4 - m}$. However, from property (iii), all $\rho \in K_{n/4}$ have the same value 1 on $n/4$ many variables, and there are only $(n^{1/2})^{3n/4} = 2^{(3/8)n \log n}$ many such ρ's. This gives a contradiction.

To prove the claim, we assume that K_μ and Q_μ have been constructed satisfying the above properties, with $\mu < n/4$. Now, we define $K_{\mu+1}$ and $Q_{\mu+1}$ inductively as follows: First, define ρ_μ by $\rho_\mu(v) = 0$ if $v \in V - Q_\mu$ and $\rho_\mu(v) = 1$ if $v \in Q_\mu$. Then, by property (ii), $||Q_\mu|| = \mu < n/4$ implies that $C\lceil_{\rho_\mu} = C_0\lceil_{\rho_\mu} = 0$ and hence $D\lceil_{\rho_\mu} = 0$. So, one of the OR gates G in D has $G\lceil_{\rho_\mu} = 0$. Fix this gate G.

Since each ρ in K_μ has $D\lceil_\rho = 1$, we have $G\lceil_\rho = 1$, and so ρ_μ differs from each $\rho \in K_\mu$ at at least one variable in G. This variable u must be in $V - Q_\mu$ and $\rho_\mu(u) = 0$ and $\rho(u) = 1$. Choose such a variable u in G that maximizes the size of the set $\{\rho \in K_\mu | \ \rho_\mu(u) = 0, \rho(u) = 1\}$. Define $Q_{\mu+1} = Q_\mu \cup \{u\}$ and $K_{\mu+1} = \{\rho \in K_\mu | \ \rho(u) = 1\}$. Then it can be verified that $K_{\mu+1}$ and $Q_{\mu+1}$ satisfy properties (i)—(iii). (In particular, G has $\le m = (1/12)n^{1/3}$ variables, and so

$$\begin{aligned} ||K_{\mu+1}|| &\ge ||K_\mu||/m \ge 2^{(n/2 - \mu/3)\log n - c\mu - m}/2^{(1/3)\log n + c} \\ &= 2^{(n/2 - (\mu+1)/3)\log n - c(\mu+1) - m}.) \end{aligned}$$

The claim, and hence the lemma, is proven. □

Theorem 4.11. There exists an oracle A such that for every $k > 0$, $BP\Sigma_k^P(A) \not\subseteq \Sigma_{k+1}^P(A)$.

Corollary 4.12. There exists an oracle A such that $MA_2(A) \subsetneq AM_2(A)$.

Proof. It is known that for all oracles A, $MA_2(A) \subseteq \Sigma_2^P(A)$ [Zachos, 1986]. □

5. Encoding and Diagonalization

In order to collapse a hierarchy to a fixed level, a common technique is to encode the information about sets in higher levels into the oracle set A so that it can be decoded using a lower level machine. If the complexity class in the higher level has a complete set,[6] then we need to encode only one set instead of infinitely many sets. For example, Baker, Gill and Solovay [1975] showed that if A is *PSPACE*-complete then $NP(A) = P(A)$ (in fact, $PSPACE(A) = P(A)$). Moreover, when we need to collapse a hierarchy to a fixed level and, using the same oracle, to separate classes below that level, the technique of encoding is used together with diagonalization, and this may create many new problems in constructing the oracle. We illustrate this technique in this section.

We first consider a simple example: collapsing NP to $NP \cap co\text{-}NP$, but keeps $P \neq NP$. That is, we need an oracle A such that $P(A) \neq NP(A) = co\text{-}NP(A)$. First, we follow the general setup in Section 3 for the diagonalization against all sets in P. Namely, we enumerate all P^1-predicate $\sigma_n(A; x)$ and choose appropriate witnesses x_n and diagonalization regions $D(n)$ so that the following requirements are to be satisfied:

$R_{n,0}$: $(\exists x_n)(\exists B \subseteq D(n))[x_n \in L_B \iff not\ \sigma_n(A(n-1) \cup B; x_n)]$,

where $L_A = \{0^{2n} \mid (\exists y, |y| = n)[0^n y \in A]\} \in NP(A)$.

In the meantime, consider a complete set $K(A)$ for $NP(A)$. For example, let $K(A) = \{\langle 0^i, a, 0^j \rangle \mid (\exists b, |b| \leq j)[\sigma_i(A; \langle a, b \rangle)$ is true and decidable in j moves$]\}$. It is important to note that whether an instance $x = \langle 0^i, a, 0^j \rangle$ is in $K(A)$ depends only on the set $A^{<|x|}$ because in j moves the machine for σ_i on input a can only query about strings in $A^{<|x|}$. To satisfy $NP(A) = co\text{-}NP(A)$, we need to make $K(A) \in co\text{-}NP(A)$, or, $\overline{K(A)} \in NP(A)$. We consider a specific $\Sigma_1^{P,1}$-predicate $\tau(A; x) \equiv (\exists y, |y| = |x|)\ 1xy \in A$, and require that for all x, $x \notin K(A) \iff \tau(A; x)$. To fit this requirement into the stage construction for requirements $R_{n,0}$, we divide it into an infinite number of requirements:

[6] A set B is complete for a class C if $B \in C$ and for all sets $C \in C$ there exists a polynomial-time computable function f such that for all x, $x \in C$ iff $f(x) \in B$.

$R_{n,1}$: $(\forall x, |x| = n)[x \notin K(A) \iff \tau(A; x)]$.

Now we describe the construction of set A. In stage n, we will satisfy requirement $R_{n,0}$ by choosing $x^n = 0^m$, where m is even, $m > t(n-1)$ and $2^{m/2} > p_n(m)$. Also in stage n, we will satisfy requirements $R_{i,1}$, for all i, $t(n-1) < 2i+1 \leq t(n)$ (again, $t(n) = p_n(m)$).

More precisely, stage n consists of four steps. In Step 1, we determine the integer m, and let $t(n) = p_n(m)$. In Step 2, we determine the memberships in A or $\overline{A}$ for strings x of length $t(n-1) < |x| < m$, and satisfy requirements $R_{i,1}$, $t(n-1) < 2i+1 < m$. This is done in $m - t(n-1) - 1$ many substeps: Substeps $t(n-1)+1, \cdots, m-1$. To begin, we let $X(t(n-1)) = A(n-1)$ and $X'(t(n-1)) = A'(n-1)$. At Substep j, where j is even, do nothing: $X(j) = X(j-1)$ and $X'(j) = X'(j-1)$. At Substep j, where j is odd, determine, for each string x of length $(j-1)/2$, whether $x \in K(X(j-1))$. If $x \in K(X(j-1))$, then let $Y_x = Y'_x = \emptyset$; otherwise, let $Y_x = \{1xy | \, |y| = |x|\}$ and $Y'_x = \emptyset$. Define $X(j) = X(j-1) \cup (\cup_{|x|=(j-1)/2} Y_x)$ and $X'(j) = X'(j-1) \cup (\cup_{|x|=(j-1)/2} Y'_x)$. Note that by the end of Substep $m-1$, we have $x \notin K(X(m-1)) \iff \tau(X(m-1); x)$ for all x, $t(n-1) < 2|x|+1 < m$, because the question of whether $x \in K(A)$ depends only on $A^{<|x|}$ and hence $x \in K(X(2|x|))$ iff $x \in K(X(m-1))$.

In Step 3, we need to satisfy $R_{n,0}$, and also make sure that requirements $R_{i,1}$, $m \leq 2i+1 \leq t(n)$, can be satisfied in Step 4. We consider the computation tree T of $\sigma_n(A; 0^m)$, with each query "$y \in ?A$" answered by $\chi_{X(m-1)}(y)$ if $|y| < m$. We also consider the following circuits:

(1) the Σ_1-circuit C_0 corresponding to the predicate "$0^m \in L_A$", and

(2) for each x, $m \leq 2|x|+1 \leq t(n)$, the circuit C_x corresponding to the predicate $\tau(A; x)$.

Let $D'(n) = \{y | \, m \leq |y| \leq t(n)\}$. To satisfy requirement $R_{n,0}$, and in the meantime, allowing requirements $R_{i,1}$, $m \leq 2i+1 \leq t(n)$, to be satisfied later, we need to find sets B_0, B_1 such that $B_0 \cup B_1 \subseteq D'(n)$, $B_0 \cap B_1 = \emptyset$, and that the following properties (a), (b) and (c) hold. Let $\rho = \rho_{B_0,B_1}$ be the restriction such that $\rho(v_z) = 1$ if $z \in B_1$, $\rho(v_z) = 0$ if $z \in B_0$, and $\rho(v_z) = *$ if $z \notin B_0 \cup B_1$.

(a) The computation tree T always accepts or always rejects, if all queries $z \in B_1$ are answered *yes* and all queries $z \in B_0$ are answered *no* (and other queries remained unanswered). We abuse the notation and write that $T\lceil_\rho \neq *$.

(b) The circuit $C_0\lceil_\rho$ is completely determined , and $C_0\lceil_\rho \neq T\lceil_\rho$.

(c) The circuits $C_x\lceil_\rho$, for all x such that $m \leq 2|x|+1 \leq t(n)$, are undetermined.

The above conditions (a) and (b) together satisfy requirement $R_{n,0}$ and condition (c) leaves $C_x\lceil_\rho$ undetermined so that all requirements $R_{i,1}$, $m \leq 2i+1 \leq t(n)$, can be satisfied in Step 4. Since variables in C_0 are those v_y's with $|y|$ being even and variables in C_x's are those v_y's with $|y| = 2|x|+1$, we can further simplify the above conditions (a), (b) and (c) into the following requirement:

R'_n: $(\exists x_n)(\exists B_0, B_1 \subseteq D'(n))[B_0 \cap B_1 = \emptyset$, $T\lceil_{\rho_{B_0,B_1}} \neq *$, and $C_0\lceil_{\rho_{B_0,B_1}} = C_x\lceil_{\rho_{B_0,B_1}} = *$ for all x such that $m \leq 2|x|+1 \leq t(n)]$.

To see that requirement R'_n can be satisfied, we let π be the fixed computation path in T in which all queries are answered *no*. Let $B_0 = \{y |\ y$ is queried in path $\pi\}$ and $B_1 = \emptyset$. Then, obviously, $T\lceil_{\rho_{B_0,B_1}} \neq *$. We note that m is chosen such that $2^{m/2} > p_n(m)$, and so $\|B_0\| \leq p_n(m) < 2^{m/2}$. Therefore, ρ_{B_0,B_1} cannot determine circuit C_0, nor any C_x, $m \leq 2|x|+1 \leq t(n)$, since each of these circuits needs at least one positive value or at least $2^{m/2}$ negative values to be completely determined.

Now, from requirement R'_n, if $T\lceil_{\rho_{B_0,B_1}} = 0$, then we choose one variable v_z in C_0 such that $z \notin B_0$, let $X(m-1) = X(m-1) \cup \{z\}$, and if $T\lceil_{\rho_{B_0,B_1}} = 1$ then do nothing. This satisfies requirement $R_{n,0}$ with respect to set $X(m-1)$.

Finally, in Step 4, we satisfy requirements $R_{i,1}$ for all i such that $m \leq 2i+1 \leq t(n)$. Again, we divide this step into $t(n)-m+1$ many substeps. In each Substep j, $j = m, \cdots, t(n)$, we consider all strings x of length $(j-1)/2$ to see whether $x \in K(X(j-1))$. If $x \in K(X(j-1))$, then let $Y_x = Y'_x = \emptyset$; otherwise, find a string $z = 1xy$, $|y| = |x|$, such that $z \notin B_0$ and let $Y_x = \{z\}$ and $Y'_x = \emptyset$ (such a string z must exist as guaranteed by condition (c) of Step 3). Define $X(j) = X(j-1) \cup (\cup_{|x|=(j-1)/2} Y_x)$ and $X'(j) = X'(j-1) \cup (\cup_{|x|=(j-1)/2} Y'_x)$. Let $A(n) = X(t(n))$ and $A'(n) = X'(t(n))$. This completes stage n.

The above stage n defined set $A(n)$, or $A^{\leq t(n)}$ if we let $A = \cup_{n=1}^{\infty} A(n)$. Note that we have, in Step 3 of stage n, satisfied requirement $R_{n,0}$ with respect to set $X(m-1)$. Note that our construction of A always keep A to be an extension of $X(j)$ and $X'(j)$ within stage n in the sense that $X(j) \subseteq A$ and $X'(j) \subseteq \overline{A}$. Therefore, requirement $R_{n,0}$ is also satisfied with respect to set A. Similarly, requirements $R_{i,1}$, for all i such that $t(n-1) < 2i+1 \leq t(n)$, are satisfied in stage n with respect to $X(2i+1)$ and $X'(2i+1)$, and so they are satisfied with respect to set A.

In summary, our construction of set A which collapses class $\mathcal{C}_3$ to class $\mathcal{C}_1$ but still separates $\mathcal{C}_1$ from $\mathcal{C}_2$ uses the following general setting:

(1) The diagonalization setting for $\mathcal{C}_1$ against $\mathcal{C}_2$ includes (a) the enumeration of predicates $\sigma_n(A;x)$ in $\mathcal{C}_2$, (b) defining a fixed set $L_A \in \mathcal{C}_1(A)$ such that the question of whether $x \in L_A$ depends only on set $A \cap W(x)$, where $W(x)$ is a window usually contained in $\{0,1\}^{|x|}$, (c) setting the requirement

$$R_{n,0}\colon (\exists x_n)[x_n \in L_A \iff \textit{not } \sigma_n(A;x_n)],$$

and (d) defining $t(n)$ such that diagonalization occurs in the region $D'(n) = \{y \mid |x_n| \leq |y| \leq t(n)\}$.

(2) To collapse the class $\mathcal{C}_3$ to $\mathcal{C}_1$, we need a complete set $K(A)$ for $\mathcal{C}_3(A)$, which has the following property: whether a string x is in $K(A)$ depends only on set $A^{<|x|}$; or, equivalently, $A^{<|x|} = B^{<|x|}$ implies that $x \in K(A) \iff x \in K(B)$. Then we need a fixed a predicate $\tau(A;x)$ in $\mathcal{C}_1$ which has the property that whether $\tau(A;x)$ holds or not depends only on set $A \cap W'(x)$, where $W'(x)$ is a window such that all windows $W'(x)$ and $W(y)$ are pairwisely disjoint. $W'(x)$ is often defined to be a subset of $\{0,1\}^{g(|x|)}$ for some function g. (In the above example, we had $W'(x) = \{1xy \mid |y| = |x|\}$, and $g(n) = 2n+1$.) The new requirements are

$$R_{n,1}\colon (\forall x, |x| = n)[x \in K(A) \iff \tau(A;x)].$$

(3) Inside the diagonalization region $D(n)$, we need to satisfy requirement $R_{n,0}$ as well as $R_{i,1}$ for all i, $t(n-1) < g(i) \leq t(n)$. Requirements $R_{i,1}$, $t(n-1) < g(i) < |x_n|$, will be done first in a straightforward way. Let $X(|x_n|-1)$ be the resulting extension of set $A(n-1)$. Then, requirement $R_{n,0}$ will be strengthened to

$$R'_n\colon (\exists B_0, B_1 \subseteq D'(n))[B_0 \cap B_1 = \emptyset \text{ and } C\lceil_{\rho_{B_0,B_1}} \neq * \text{ and } C_0\lceil_{\rho_{B_0,B_1}} = C_x\lceil_{\rho_{B_0,B_1}} \neq *, \text{ for all } x,\ |x_n| \leq g(|x|) \leq t(n)],$$

where C_0 is the circuit corresponding to the predicate "$x_n \in L_A$", C_x is the circuit corresponding to the predicate $\tau(A;x)$, and C is the circuit corresponding to the predicate $\sigma_n(A;x_n)$, with the variables v_y replaced by the value $\chi_{X(|x_n|-1)}(y)$, if $|y| < |x_n|$, and ρ_{B_0,B_1} is the restriction defined above. Observe that the variables in C_0 are those v_y's such that $y \in W(x_n)$, and the variables in C_x are those v_y's such that $y \in W'(x)$. So, all circuits except circuit C have pairwisely disjoint variables. This verifies that R'_n implies $R_{n,0}$. Also note that now $R_{n,0}$ is, again, reduced to a lower bound requirement on circuits—this time a little more complex than the lower bound requirements defined in Sections 3 and 4.

Finally, when R'_n is satisfied, requirements $R_{i,1}$, $|x_n| \leq g(i) \leq t(n)$, can be satisfied since we have left $C_x\lceil_{\rho_{B_0,B_1}}$ undertimed for all x of length i.

6. Collapsing Results

In this section, we demonstrate how to apply the general setting of Section 5 to collapse some hierarchy to a fixed level, while keeping other classes separated. This task is generally more complex than the separation results. In particular, when the classes to be separated are not of an apparently simpler structure than the classes to be collapsed, some different types of encoding techniques must be used. These special techniques are presented in Sections 6.2 and 6.3.

6.1. Collapsing PSPACE to PH

We first construct an oracle A such that $PSPACE(A) = \Sigma_k^P(A) \neq \Sigma_{k-1}^P(A)$, where k is an arbitrary but fixed integer greater than 0.

Following the setting given in Section 5, we let

$$L_A = \{0^{(k+1)n} \mid (\exists y_1, |y_1| = n) \cdots (Q_k y_k, |y_k| = n)\ 0^n y_1 \cdots y_k \in A\},$$

$$\tau(A; x) \equiv (\exists y_1, |y_1| = |x|) \cdots (Q_k y_k, |y_k| = |x|)\ [1xy_1 \cdots y_k \in A], \text{ and}$$

$$Q(A) = \{\langle 0^i, a, 0^j\rangle \mid \text{the } i\text{th TM } M_i \text{ accepts } a \text{ using } \leq j \text{ cells}\}.$$

In the above definition for $Q(A)$, the space used by the query tape of machine M_i is included in the space measure. Therefore, set $Q(A)$ is complete for $PSPACE(A)$, and whether x is in $Q(A)$ depends only on set $A^{<|x|}$. In addition, let $W(0^m) = \{0^{m/(k+1)}z \mid |z| = km/(k+1)\}$ and $W'(x) = \{1xz \mid |z| = k|x|\}$. Then, the question of whether $0^m \in L_A$ depends only on set $A \cap W(0^m)$ and the predicate $\tau(A; x)$ depends only on set $A \cap W'(x)$. Also, all windows $W(0^m)$ and $W'(x)$ are pairwisely disjoint. Let $g(i) = (k+1)i + 1$.

In stage n, we choose a sufficiently large m such that m is a multiple of $k+1$ and is greater than $t(n-1)$. Let $x_n = 0^m$ and $t(n) = p_n(m)$. As in Section 5, we first perform Steps 1 and 2 and define set $X(n-1)$. Then, let C be the circuit corresponding to the nth $\Sigma_{k-1}^{P,1}$-predicate $\sigma_n(A; x_n)$, with the variables v_y replaced by the value $\chi_{X(m-1)}(y)$, if $|y| < m$, C_0 be the circuit corresponding to the predicate "$0^m \in L_A$", and C_x be the circuit corresponding to the predicate $\tau(A; x)$. From the discussion in Section 5, our construction in stage n is reduced to the following requirement (if $k \geq 2$):

R'_n: $(\exists B_0, B_1 \subseteq D'(n))[B_0 \cap B_1 = \emptyset$ and $C\lceil_{\rho_{B_0,B_1}} \neq *$ and $C_0\lceil_{\rho_{B_0,B_1}} = C_x\lceil_{\rho_{B_0,B_1}} \neq *$, for all x, $m \leq g(|x|) \leq t(n)]$,

where ρ_{B_0,B_1} is the restriction defined by making all v_z to be 1 if $z \in B_1$ and all v_z to be 0 if $z \in B_0$. (Note that when $k = 1$, the construction is almost identical to that given in Section 5, with the *NP*-complete set $K(A)$ replaced by the *PSPACE*-complete set $Q(A)$.)

It is clear that C is a $\Sigma_{k-1}(p_n(m))$-circuit, while each of C_0 and C_x contains a subcircuit computing a function $f_k^{2^{m/(k+1)}}$. This allows us to show that requirement R'_n can be satisfied by showing the following generalization of the lower bound results of Theorem 4.6.

Theorem 6.1. Let $\{C_i\}_{i=1}^t$ be t circuits, each computing a function f_k^n, with pairwisely disjoint variables. Let C be a Σ_{k-1}-circuit having size $\leq r$ and bottom fanin $\leq \log r$. If $t \leq 2^{\delta n^{1/6}}$ and $r \leq 2^{\delta n^{1/3}}$, with $\delta = 1/12$, then for sufficiently large n, there exists a restriction ρ such that $C\lceil_\rho \neq *$ and $C_i\lceil_\rho = *$ for all $i = 1, \cdots, t$.

Proof. The proof is very similar to that of Theorem 4.6. First, for the purpose of induction, we change the induction statement to a stronger form that the theorem holds if C satisfies the weaker constraint that it has at most r gates not at the bottom level.

The base step of the induction proof, with $k = 2$, is trivial. We leave it to the reader. For the inductive step, we define probability spaces $R^+_{q,\mathcal{B}}$ and $R^-_{q,\mathcal{B}}$ as in the proof of Theorem 4.6, where $\mathcal{B}$ is the partition of variables such that all variables leading to a bottom gate in any C_i form a block. We need to show

(a) for a random restriction ρ, the probability that $C\lceil_\rho$ is equivalent to a Σ_{k-2}-circuit having $\leq r$ gates not at the bottom level and having bottom fanin $\leq \log r$ is big, and

(b) for a random restriction ρ, the probability that each $C_i\lceil_\rho$, $1 \leq i \leq t$, contains a subcircuit computing a function f_{k-1}^n is big.

Part (a) follows exactly from the proof of Theorem 4.6. Part (b) can be proved by a slight modification of the proof of Lemma 4.8. Namely, we let q_1 be the probability that bottom AND gates $H_j\lceil_{\rho g(\rho)}$ of $C_i\lceil_{\rho g(\rho)}$, $1 \leq i \leq t$, takes the value s_j, and let q_2 be the probability that all OR gates at level 2 from bottom of all

$C_i\lceil_{\rho g(\rho)}$, $1 \le i \le t$, have $\ge \sqrt{n}$ many child nodes $H_j\lceil_{\rho g(\rho)}$ having value $s_j = *$. Note that in the proof of Lemma 4.8, the estimation for q_1 was

$$q_1 \ge (1 - e^{-n^{1/6}})^{n^{k-1}},$$

because there were n^{k-1} many bottom gates in C_0. Here, we have t such circuits, each having n^{k-1} many bottom gates, so

$$q_1 \ge (1 - e^{-n^{1/6}})^{n^{k-1} \cdot t} \ge 1 - e^{-n^{1/6}} \cdot n^{k-1} \cdot 2^{\delta n^{1/6}} \ge 5/6,$$

if n is sufficiently large.

Similarly, we modify the estimation for probability q_2 and obtain

$$q_2 \ge (1 - 2^{-n^{1/2}})^{n^{k-2} \cdot t} > 5/6,$$

for sufficiently large n. □

Theorem 6.2. For every $k \ge 1$, there exists a set A such that $PSPACE(A) = \Sigma_k^P(A) \ne \Sigma_{k-1}^P(A)$.

Proof. There are $\le 2^{t(n)} = 2^{p_n(m)}$ many circuits C_x, each having a subcircuit computing a function $f_k^{2^{m/(k+1)}}$. To satisfy requirement R'_n, all we need is to choose m so large that Theorem 6.1 holds with respect to the function $f_k^{2^{m/(k+1)}}$ and that $(1/12)2^{m/6(k+1)} > k \cdot p_n(m)$. □

6.2. Collapsing PH but Keeping PSPACE Seperated from PH

In this section, we construct an oracle A such that $PSPACE(A) \ne \Sigma_{k+1}^P(A) = \Sigma_k^P(A) \ne \Sigma_{k-1}^P(A)$ for an arbitrary $k \ge 1$.

The separation part consists of two requirements: $PSPACE(A) \ne \Sigma_k^P(A)$ and $\Sigma_k^P(A) \ne \Sigma_{k-1}^P(A)$. Let

$$L_A = \{0^m \mid \|A^{=n}\| \text{ is odd}\} \in PSPACE(A),$$

and

$$L'_A = \{0^{(k+1)n} \mid (\exists y_1, |y_1| = n) \cdots (Q_k y_k, |y_k| = n) 0^n y_1 \cdots y_k \in A\} \in \Sigma_k^P(A).$$

These requirements may be divided into an infinite number of requirements:

$R_{2n,0}$: $(\exists x_{2n})[x_{2n} \in L_A \iff not\ \sigma_n^{(k)}(A; x_{2n})]$,

$R_{2n+1,0}$: $(\exists x_{2n+1})[x_{2n+1} \in L'_A \iff not\ \sigma_n^{(k-1)}(A; x_{2n+1})]$,

where $\sigma_n^{(h)}$ is the nth $\Sigma_h^{P,1}$-predicate.

The collapsing part requires that $\Sigma_{k+1}^P(A) = \Sigma_k^P(A)$. We assume the existence of a $\Sigma_{k+1}^P(A)$-complete set $K^{k+1}(A)$ which has the property that the question of whether $x \in K^{k+1}(A)$ depends only on set $A^{<|x|}$. Next we let $\tau(A;x)$ be a fixed $\Sigma_k^{P,1}$-predicate:

$$\tau(A;x) \equiv (\exists y_1, |y_1| = |x|) \cdots (Q_k y_k, |y_k| = |x|)\ 1xy_1 \cdots y_k \in A.$$

We divide the requirement of $\Sigma_{k+1}^P(A) = \Sigma_k^P(A)$ into the following requirements:

$R_{i,1}$: $(\forall x, |x| = i)$ $[x \in K^{k+1}(A) \iff \tau(A;x)]$.

We now describe the construction of set A. In stage $2n+1$, we will satisfy requirements $R_{2n+1,0}$, as well as requirements $R_{i,1}$, for all i such that $t(n-1) < g(i) \leq t(n)$, where $t(n-1)$ and $t(n)$ are the bounds for the diagonalization region described in Sections 3 and 5, and $g(i) = (k+1)i + 1$. The construction is almost identical to the stage n of the construction in Section 6.1. The only difference is that we now are working with the complete set $K^{k+1}(A)$ instead of the $PSPACE(A)$-complete set $Q(A)$.

In stage $2n$, we will satisfy requirement $R_{2n,0}$, as well as requirements $R_{i,1}$, for all i such that $t(n-1) < g(i) \leq t(n)$. Following the setting of Section 5, we let $x_{2n} = 0^m$, with m being a multiple of $k+1$ and is sufficiently large. We now consider the following circuits:

(a) the circuit C corresponding to the predicate $\sigma_n^{(k)}(A; 0^m)$, with all variables v_y replaced by $\chi_{A(n-1)}(y)$ if $|y| < t(n-1)$,

(b) for each x, $t(n-1) < g(|x|) \leq t(n)$, the circuit C_x corresponding to the predicate $\tau(A;x)$.

We need a restriction ρ on variables such that $C\lceil_\rho \neq *$, $C_x\lceil_\rho = *$ for all x, $t(n-1) < g(|x|) \leq t(n)$, and the set $\{y \mid |y| = m, \rho(v_y) = *\}$ is nonempty. (Note that none of the variables v_y in C_x's is of length m.) Inspecting the structure of these circuits, we learn that C is a $\Sigma_k(p_n(m))$-circuit and each C_x is a $\Sigma_k(|x|)$-circuit. Thus, such a restriction ρ does not seem to exist (e.g., circuit C may simulate a particular C_x). Our solution to this problem is to modify these requirements so that the requirements $R_{i,1}$, even for $g(i) \geq m$, can be satisfied *before* the requirement $R_{2n,0}$ is satisfied. The price we pay is that the circuit C would become more complicated—though still not complicated enough to be able to compute the parity of $A^{=m}$.

We now describe how this tradeoff between the structure of circuits C and C_x's is done. First, we replace each requirement $R_{i,1}$ by a simpler requirement $R'_{i,1}$

(only for those i such that $t(n-1) < g(i) \le t(n)$):

$R'_{i,1}$: $(\forall x, |x| = i)[x \in K^{k+1}(A) \Rightarrow (\forall z, |z| = ki) 1xz \in A]$ and $[x \notin K^{k+1}(A) \Rightarrow (\forall z, |z| = ki) 1xz \notin A]$.

Using these simpler requirements, we can modify circuit C to depend only on variables v_y, $|y| = m$, and so the requirement $R_{2n,0}$ may be satisfied without worrying about requirements $R_{i,1}$. The modification will make circuit C having depth greater than $k+1$, but within the acceptable size. Let $V = \{v_y | \ v_y$ occurs in C and $y = 1xz$ for some x and z, $|z| = k|x|$, and $|y| > t(n-1)\}$. For each x of length $> (t(n-1)-1)/(k+1)$, let D_x be the circuit corresponding to the predicate $x \in K^{k+1}(A)$, with its variables v_y replaced by $\chi_{A(n-1)}(y)$ if $|y| < t(n-1)$, and replaced by 0 if $|y| \neq m$ and $v_y \notin V$. Note that D_x is a Σ_{k+1}-circuit such that all its variables have the form $v_{x'}$, with $|x'| < |x|$, because the question of "$x \in ?K^{k+1}(A)$" depends only on set $A^{<|x|}$. In addition, we note that for every set A, if A satisfies

$$(*) \qquad A^{\le t(n-1)} = A(n-1), A \cap \{y | \ |y| > t(n-1), |y| \neq m, v_y \notin V\} = \emptyset$$

then $D_x \lceil_{\rho_A} = 1$ iff $x \in K^{k+1}(A)$.

Now we modify circuit C as follows. First, for all variables v_y with $|y| \le t(n-1)$, replace v_y by $\chi_{A(n-1)}(y)$. Then, for all variables $v_y \notin V$ such that $|y| \neq m$, replace v_y by constant 0. Second, for all variables $v_y \in V$, if $y = 1xz$ for some x and z, $|z| = k|x|$, then we replace it by circuit D_x (and replace $\overline{v}_y$ by the dual circuit of D_x). Note that after the modification, we obtain a circuit C_1 of depth $\le 2(k+1)$ but each of its variables v_y is either of length $|y| = m$ or is in V and of length $|y| < p_n(m)/(k+1)$. Also note that if we apply a restriction ρ_A to circuit C_1, then $C_1 \lceil_{\rho_A} = 1$ iff $\sigma_n(A; 0^m)$, provided that requirements $R'_{i,1}$ are satisfied by A for all i such that $t(n-1) < g(i) \le t(i)$, and that A satisfies $(*)$. This is true because requirements $R'_{i,1}$ imply that for each $y = 1xz$, $|z| = k|x|$, $x \in K^{k+1}(A) \iff y \in A$ and so $D_x \lceil_{\rho_A} = 1 \iff y \in A$.

Repeat the above process until the circuit no longer has any variable in V (thus, all variables v_y have length $|y| = m$). To obtain such a circuit C', we need only to repeat the above modification for at most $\log(p_n(m))$ times. Therefore, the resulting circuit C' has depth $\le \log(p_n(m)) \cdot (k+1)$, and has fanins $\le 2^{q(p_n(m))}$, where q is a polynomial depending on the set $K^{k+1}(A)$ (i.e., $2^{q(|x|)}$ bounds the fanins of circuit D_x). Also, if A satisfies $(*)$ and requirements $R'_{i,1}$ for all i such that $t(n-1) < g(i) \le t(i)$, then $C' \lceil_{\rho_A} = 1$ iff $\sigma_n(A; 0^m)$. Since C' does not have common variables with C_x's, we need only to show that C' does not compute the parity of 2^m variables. This, again, reduces our diagonalization problem to the lower bound problem about parity circuits.

Once it is proved that this modified circuit C' does not compute the parity of set $A^{=m}$, we can complete the stage $2n$ by finding a subset $B \subseteq A^{=m}$ such that ρ_B completely determines C' but $C'\lceil_{\rho_B} = 1$ iff $\|B\|$ is even. Then, we satisfy requirements $R'_{i,1}$ for each i, $t(n-1) < g(i) \le t(n)$, by letting each $y = 1xz$, $|z| = k|x|$, be in $A(n)$ iff $x \in K^{k+1}(A)$ for all x, $|x| = i$. (More precisely, we do this in $t(n) - t(n-1)$ substeps. We let $X(t(n-1)) = A(n-1)$ and, in each Substep j, $t(n-1) < j \le t(n)$, we let strings $y = 1xz$ be in $X(j)$ iff $x \in K^{k+1}(X(j-1))$ for all x, $g(|x|) = j$. Finally, let $A(n) = X(t(n))$.) Note that set $A(n)$ satisfies both (*) and $R'_{i,1}$ for all i, $t(n-1) < g(i) \le t(n)$. It is left to show that circuit C' does not compute the parity of set $A^{=m}$.

Theorem 6.3. Let $s(n)$ be the minimum size of circuit C of depth $k = c \log\log n$ for some $c > 0$ such that C computes the parity of n variables. Then, for sufficiently large n, $s(n) \ge 2^{\epsilon n^{1/(k-1)}}$ for some $\epsilon > 0$.

Proof. We proved in Theorem 4.1 and Remark 4.3 that for some $\epsilon > 0$, $s_k(n) \ge 2^{\epsilon n^{1/(k-1)}}$ for depth-k parity circuits, if $n > n_k$ for some constant n_k depending only on k. In fact, the constant n_k can be taken as $(n_0)^k$ for some absolute constant n_0 which does not depend on k (see Hastad [1987] for details). So, if $n > (n_0)^{c \log\log n}$, then $s(n) \ge 2^{\epsilon n^{1/(k-1)}}$. □

Theorem 6.4. For each $k \ge 1$, there exists an oracle A such that $PSPACE(A) \ne \Sigma_{k+1}^P(A) = \Sigma_k^P(A) \ne \Sigma_{k-1}^P(A)$. Also, there exists an oracle B such that $PSPACE(B) \ne NP(B) = P(B)$.

The above proof actually showed that the set $L_A = \{0^m \mid \|A^{=m}\| \text{ is odd}\}$ is not in $PH(A)$ even if $PH(A)$ is finite. Therefore, for this set A, $\oplus P(A) \not\subseteq PH(A)$. In fact, all we need above is that L_A is not in $\Sigma_{c \log n}^P(A)$ for all $c > 0$. Thus, the same proof established the following more general result.

Corollary 6.5. If $\mathcal{C}$ is a complexity class such that $\mathcal{C}(A) \not\subseteq \Sigma_{c \log n}^P(A)$ for some oracle A, then for every $k > 0$, there exists an oracle B such that $\mathcal{C}(B) \not\subseteq \Sigma_{k+1}^P(B) = \Sigma_k^P(B) \ne \Sigma_{k-1}^P(B)$.

It is an interesting question whether there exists an oracle A such that the class $\Sigma_{\log n}^P(A)$ is outside the polynomial hierarchy $PH(A)$, and $PH(A)$ collapses to the kth level for some fixed but arbitrary k.

6.3. Collapsing BPH to PH but Keeping PH Infinite

In Section 4.3, we have shown that there exists an oracle A such that $BP\Sigma_k^P(A) \not\subseteq \Sigma_{k+1}^P(A)$ for all $k \geq 1$, and hence both hierarchies $PH(A)$ and $BPH(A)$ are infinite and the two differ at each level. What we want to show now is that relative to some oracle A the two hierarchies are identical in the sense that $BP\Sigma_k^P(A) = \Sigma_k^P(A)$ for all $k \geq 0$ and yet $PH(A)$ is infinite. Note that $\Sigma_k^P(A) \subseteq BP\Sigma_k^P(A) \subseteq \Pi_{k+1}^P(A)$ for all oracles A. Therefore our proof needs to collapse $BP\Sigma_k^P(A)$ to $\Sigma_k^P(A)$ but keeping $\Pi_{k+1}^P(A)$ separated from $\Sigma_k^P(A)$. Such a proof, like the one in Section 6.2, does not fit into the general setting of Section 5 (where the setting is designed for separating classes which lie below in the hierarchy than the classes to be collapsed).

The reader who is familiar with the theory of relativization would notice that the intended result here is a generalization of Rackoff's [1982] result that there exists an oracle A such that $P(A) = R(A) \neq NP(A)$. Rackoff's result also follows from Bennett and Gill's [1981] proof that for a random oracle A, $P(A) = R(A)$ and $P(A) \neq NP(A)$. We remark that neither of these proofs seems to work for our generalized result. Rackoff's constructive proof requires the oracle to be a sparse set, but from Balcázar, Book and Schöning [1986] and Long and Selman [1986], a sparse set does not seem to be able to separate Σ_2^P from Π_2^P (unless $\Sigma_2^P \neq \Pi_2^P$ in the unrelativized form). For the approach of using random oracles, we can see from Bennett and Gill's work that $BP\Sigma_k^P(A) = \Sigma_k^P(A)$ for random oracles A. However, it is still an important open question whether the polynomial hierarchy is infinite relative to a random oracle (cf. Cai [1986], Babai [1987] and Hastad [1987]).

First we simplify our problem to only collapsing $BP\Sigma_k^P(A)$ to $\Sigma_k^P(A)$ while keeping $\Sigma_k^P(A) \neq \Sigma_{k+1}^P(A)$, for a fixed but arbitrary $k > 0$. The separation part can be handled by a usual diagonalization setting. Let $L_A = \{0^{(k+2)n} \mid (\exists y_1, |y_1| = n) \cdots (Q_{k+1} y_{k+1}, |y_{k+1}| = n)\ 0^n y_1 \cdots y_{k+1} \in A\} \in \Sigma_{k+1}^P(A)$. Let $\sigma_n(A; x)$ be an enumeration of all $\Sigma_k^{P,1}$-predicates. Our requirements for separation are

$R_{n,0}$: $(\exists x^n = 0^m)[0^m \in L_A \iff \mathit{not}\ \sigma_n(A; 0^m)]$.

For the collapsing part, first recall that $BP\Sigma_k^P(A)$ has a simple characterization (see, for example, Zachos [1986]): If $k \geq 1$ then $L \in BP\Sigma_k^P(A)$ iff there exists a $\Sigma_k^{P,1}$-predicate σ such that for all x,

$$x \in L \Rightarrow \forall_{\mathrm{p}} y\ \sigma(A; \langle x, y\rangle),$$

$$x \notin L \Rightarrow \exists_{\mathrm{p}}^{+} y\ \mathit{not}\ \sigma(A; \langle x, y\rangle).$$

We will use this characterization in the following proof.

One of the difficulty in setting the requirements for the collapsing result is that the class $BP\Sigma_k^P(A)$ is not known to possess a complete set. Therefore, the encoding of information becomes more complicated. Fortunately, we can find a pair of *pseudo-complete* sets for $BP\Sigma_k^P(A)$:

$$J_1(A) = \{\langle 0^i, a, 0^j\rangle \mid j \geq p_i(|a|), (\forall b, |b| = j)\ \sigma_i(A; \langle a, b\rangle)\},$$

$$J_0(A) = \{\langle 0^i, a, 0^j\rangle \mid j \geq p_i(|a|), (\exists\ (3/4)\cdot 2^j \text{ many } b, |b| = j)\ \textit{not}\ \sigma_i(A; \langle a, b\rangle)\},$$

where σ_i is the ith $\Sigma_k^{P,1}$-predicate. (We assume further that if $\sigma_i(A; \langle a, b\rangle)$ and b is an initial segment of b' then $\sigma_i(A; \langle a, b'\rangle)$.) From the extra condition that $j \geq p_i(|a|)$, we can see that the question of whether $x \in J_1(A)$ or $x \in J_0(A)$ depends only on set $A^{<|x|}$. The "reduction" from a set $L \in BP\Sigma_k^P(A)$ to the pair $(J_1(A), J_0(A))$ is easy to see from the characterization for $BP\Sigma_k^P(A)$: For each $L \in BP\Sigma_k^P(A)$, there exists an i such that $x \in L \Rightarrow \langle 0^i, x, 0^{p_i(|x|)}\rangle \in J_1(A)$ and $x \notin L \Rightarrow \langle 0^i, x, 0^{p_i(|x|)}\rangle \in J_0(A)$. This allows us to set up our requirements as

$R_{n,1}$: $(\forall x = \langle 0^i, a, 0^j\rangle, |x| = n)[x \in J_1(A) \Rightarrow \tau(A; x)$ and $x \in J_0(A) \Rightarrow$ *not* $\tau(A; x)$],

where τ is a fixed $\Sigma_k^{P,1}$-predicate to be defined as follows.

It is natural to try to use an arbitrary $\Sigma_k^{P,1}$-predicate for $\tau(A; x)$; e.g., the one used in Section 6.1: $\tau(A; x) \equiv (\exists y_1, |y_1| = |x|)\cdots(Q_k y_k, |y_k| = |x|)\ 1xy_1\cdots y_k \in A$. Unfortunately, if we use such a predicate τ, then the conflict between the separating requirements and the encoding requirements would be too much to overcome. Instead, we let $\tau(A; x)$ be a simulation of $\sigma_i(A; \langle a, b\rangle)$ for "random" choices of b (if $x = \langle 0^i, a, 0^j\rangle$). Namely, for each n and i, $0 \leq i \leq 2^n - 1$, let $s_{n,i}$ be the ith string in $\{0,1\}^n$, under the lexicographic order, and for each n and r, $1 \leq r \leq n$, let $w_{n,r}^A = \chi_A(s_{n,(r-1)n})\cdots\chi_A(s_{n,rn-1})$, i.e., $w_{n,1}^A, \cdots, w_{n,n}^A$ are n n-bit strings determined by set $A^{=n}$. Now let $\tau_i(A; x)$ be true iff $\sigma_i(A; \langle a, w_{n,r}^A\rangle)$ are true for all r, $1 \leq r \leq n$, where $n = 2|x|$. It is clear that for all x, $x \in J_1(A) \Rightarrow \tau(A; x)$; that is, requirement $R_{n,1}$ is simplified to $(\forall x, |x| = n)$ $[x \in J_0(A) \Rightarrow \textit{not}\ \tau(A; x)]$.

At stage n, we try to satisfy requirement $R_{n,0}$, as well as requirements $R_{j,1}$ for j such that $t(n-1) < 2j \leq t(n)$. The critical step is, of course, to satisfy $R_{n,0}$ and still keeps predicates $\tau(A; x)$, $t(n-1) < 2|x| \leq t(n)$, undetermined. We choose $x_n = 0^m$, where m is a sufficiently large odd integer greater than $t(n-1)$. Also, we let $t(n) = p_n(m)$. Then, we satisfy all requirements $R_{j,1}$ for all j such that

$t(n-1) < 2j < m$, by set $X(m-1)$, which is an extension of $A(n-1)$. The existence of such an extension is nontrivial, but will become clear later (see Remark 6.8).

Next, let C_0 be the Σ_{k+1}-circuit corresponding to the predicate "$0^m \in L_A$". Then, C_0 has a subcircuit computing a function $f_{k+1}^{2^{m/(k+2)}}$. Let C be the $\Sigma_k(p_n(m))$-predicate corresponding to the predicate $\sigma_n(A; 0^m)$, with all variables v_y, $|y| < m$, replaced by $\chi_{X(m-1)}(y)$. In order to perform diagonalization while leaving predicates $\tau(A; x)$ undetermined, we must modify circuit C as we did in Section 6.2. However, we cannot increase its depth here, because C_0 is barely one level deeper than C. What we will do is to simulate *all* possible answers for queries "$y \in ? A$" asked by $\sigma_n(A; 0^m)$ if $y = s_{i,j}$, for some even integer i, $m \le i \le p_n(m)$, and for some j, $0 \le j \le i^2 - 1$.

Let $h(\ell) = \sum\{i^2 | \ \ell \le i \le t(n), i \text{ even}\}$. We identify each string α of length $h(\ell)$ with a subset B_α of $E_\ell = \{s_{i,j} | \ \ell \le i \le t(n), i \text{ even}, \ 0 \le j \le i^2 - 1\}$ such that $\alpha = \chi_{B_\alpha}(s_{\ell,0}) \cdots \chi_{B_\alpha}(s_{\ell,\ell^2-1})\chi_{B_\alpha}(s_{\ell+1,0}) \cdots \chi_{B_\alpha}(s_{t(n),t(n)^2-1})$ (assuming that both ℓ and $t(n)$ are even). Now, for each string α of length $h(m+1)$, let C_α be the circuit modified from C by (a) replacing all variables v_y, $y = s_{i,j} \in E_{m+1}$, by $\chi_{B_\alpha}(s_{i,j})$, and (b) replacing all variables v_y, $|y| \ne m$ and $y \notin E_{m+1}$, by constant 0. Then, each C_α contains only variables v_y of length $|y| = m$. We are ready to reduce our requirements $R_{n,0}$ and $R_{j,1}$'s to the lower bound problem on circuits C_0 versus C_α's. More precisely, we need the following lemma (but postpone the proof).

Lemma 6.6. Let $k \ge 1$ and h and q be two polynomial functions. Let C_α, $1 \le \alpha \le 2^{h(m+1)}$, be $2^{h(m+1)}$ many $\Sigma_k(q(m))$-circuits. Let C_0 be a circuit computing a function $f_{k+1}^{2^{m/(k+2)}}$. Then, for sufficiently large m, there exists an assignment ρ on variables such that $\|\{\alpha | \ C_\alpha\lceil_\rho \ne C_0\lceil_\rho\}\| \ge 2^{h(m+1)-(m+1)}$.

To see why this lemma is sufficient for our requirements, let us assume that such an assignment ρ has been chosen , and we let $X(m) = X(m-1) \cup \{y | \ |y| = m, \rho(v_y) = 1\}$. Also let $T = \{\alpha | \ C_\alpha\lceil_\rho \ne C_0\lceil_\rho\}$. Note that $\|T\| \ge 2^{h(m+1)-(m+1)}$. We claim that we can find a subset $B_{m+1} \subseteq \{s_{m+1,0}, \cdots, s_{m+1,(m+1)^2-1}\}$ such that

(a) $R_{(m+1)/2,1}$ is satisfied by $X(m) \cup B_{m+1}$, and

(b) the corresponding string $\beta_{m+1} = \chi_{B_{m+1}}(s_{m+1,0}) \cdots \chi_{B_{m+1}}(s_{m+1,(m+1)^2-1})$ has the property that the set $T_{\beta_{m+1}} =_{\text{defn}} \{\gamma \in \{0,1\}^{h(m+3)} | \ \beta_{m+1}\gamma \in T\}$ has size $\|T_{\beta_{m+1}}\| \ge 2^{h(m+3)-(m+2)}$.

To prove this claim, we first state a combinatorial lemma.

Lemma 6.7. Let $\ell < 2^{(m+1)/2}$. Let D be a $\ell \times 2^{m+1}$ boolean matrix such that each row of D has $\geq (3/4) \cdot 2^{m+1}$ many 1's. Then, for a randomly chosen $(m+1)$-tuple $(j_1, \cdots, j_{m+1})$, $1 \leq j_r \leq 2^{m+1}$,

$$\Pr[(\forall i, 1 \leq i \leq \ell)(\exists r, 1 \leq r \leq m+1) D[i, j_r] = 1] \geq 1 - 2^{-(m+3)}.$$

Now we form the matrix D as follows: each row is labeled by a string x of length $(m+1)/2$ and $x \in J_0(X(m))$, and each column is labeled by a string $z \in \{0,1\}^{m+1}$. For each x and z, let $D[x,z] = 1$ iff $[x = \langle 0^i, a, 0^j \rangle$ and $\sigma_i(X(m); \langle a, z\rangle)$ is false]. Then, D satisfies the hypothesis of Lemma 6.7 and hence the conclusion. That is, the set $S_{m+1} = \{z_1 \cdots z_{m+1} | \; |z_1| = \cdots = |z_{m+1}| = m+1, (\forall x \in J_0(X(m)) \cap \{0,1\}^{(m+1)/2})(\exists j, 1 \leq j \leq m+1) D[x, z_j] = 1\}$ has size $\|S_{m+1}\| \geq (1 - 2^{-(m+3)}) 2^{(m+1)^2}$. Thus, for any $\beta_{m+1} \in S_{m+1}$, $R_{(m+1)/2,1}$ is satisfied by set $X(m) \cup B_{m+1}$ for each corresponding set B_{m+1} (i.e., $s_{m+1,j} \in B_{m+1}$ iff the jth bit of β_{m+1} is 1).

Now, it takes a simple counting argument to see that there exists a $\beta_{m+1} \in S_{m+1}$ such that $\|T_{\beta_{m+1}}\| \geq 2^{h(m+3)-(m+2)}$. This proves the claim. □

We let $X(m+1) = X(m) \cup B_{m+1}$. The above showed that we can satisfy requirement $R_{m+1,1}$ and yet keep many "good" strings α. It can be checked that the above process can be repeated for all the requirements $R_{j,1}$, $m+1 \leq 2j \leq t(n)$, and keeping size $\|T_{\beta_{2j}}\| \geq 2^{h(2j+2)-(2j+1)}$. In particular, $\|T_{\beta_{t(n)-2}}\| \geq 2^{t(n)^2 - (t(n)-1)}$, and we only need to choose a string $\beta_{t(n)} \in T_{\beta_{t(n)-2}} \cap S_{t(n)}$ (where $S_{t(n)}$ is defined similar to S_{m+1}, having size $\|S_{t(n)}\| \geq (1 - 2^{-(t(n)+2)}) 2^{t(n)^2}$). Let $\beta = \beta_{m+1}\beta_{m+3} \cdots \beta_{t(n)}$ and define $X(t(n))$ accordingly. We note that $\beta \in T$ implies that $C_0 \lceil_\rho \neq C_\beta \lceil_\rho$ for some assignment ρ on variables v_y, $|y| = m$. Since C_β is the circuit C with all variables v_y, $|y| \neq m$, replaced by $\chi_{X(t(n))}(y)$, we see that requirement $R_{n,0}$ is satisfied by set $X(t(n))$. Let $A(n) = X(t(n))$. This completes the stage n.

Remark 6.8. At this moment, we observe that earlier in stage n, the construction of set $X(m-1)$ can be done just like the above construction of set $X(t(n))$. Actually, we don't need part (b) of the claim, and hence it is easier.

The only thing left to show is Lemma 6.6.

Proof of Lemma 6.6. We prove it by induction on k.

First consider the case $k = 1$. We show that the lemma holds even if C_0 is an AND of $2^{m/2(k+2)}$ variables. Let $r = m/2(k+2)$. We let the 2^r variables be $v_1, \cdots, v_{2^r}$, and define $2^r + 1$ assignments ρ_i, $0 \leq i \leq 2^r$, as follows: $\rho_0(v_j) = 1$ for all

j; and for each $i \geq 1$, $\rho_i(v_j) = 1$ if $j \neq i$ and $\rho_i(v_j) = 0$ if $j = i$. Note that $C_0\lceil_{\rho_0} = 1$ and $C_0\lceil_{\rho_i} = 0$ for all $i \geq 1$. For each i, $0 \leq i \leq 2^r$, let

$$E_i = \{\alpha \mid 1 \leq \alpha \leq 2^{h(m+1)}, C_\alpha\lceil_{\rho_i} = C_0\lceil_{\rho_i}\}.$$

Suppose, by way of contradiction, that the lemma does not hold for a specific C_0. Then each E_i has size $\|E_i\| \geq (1 - 2^{-(m+1)})2^{h(m+1)}$ and hence the intersection of all E_i's must be nonempty. Let β be a specific string in the intersection of E_i's. We have $C_\beta\lceil_{\rho_i} = C_0\lceil_{\rho_i}$ for all i, $0 \leq i \leq 2^r$. Note that $C_\beta\lceil_{\rho_0} = 1$ implies that C_β has a subcircuit D having $D\lceil_{\rho_0} = 1$. However, this subcircuit D is an AND of only $q(m)$ many inputs. Assume that m is so large that $2^r > q(m)$. There must be some v_j such that neither v_j nor $\overline{v}_j$ occurs in D, and hence

$$D\lceil_{\rho_0} = 1 \Rightarrow D\lceil_{\rho_j} = 1 \Rightarrow C_\beta\lceil_{\rho_j} = 1,$$

which is a contradiction.

For the inductive step, let $k > 1$. From the proof of Theorem 6.1, we can find a restriction ρ such that $C_0\lceil_\rho$ contains a subcircuit computing a $f_k^{2^{m/(k+2)}}$ function, and for each α, $C_\alpha\lceil_\rho$ is a $\Sigma_{k-1}(q(m))$-circuit. By the inductive hypothesis, there exists an assignment ρ' such that $\|\{\alpha \mid C_\alpha\lceil_{\rho\rho'} \neq C_0\lceil_{\rho\rho'}\}\| \geq 2^{h(m+1)-(m+1)}$. The combined restriction $\rho\rho'$ satisfies our requirement. □

Theorem 6.9. For every $k > 0$, there exists an oracle A such that $\Sigma_k^P(A) = BP\Sigma_k^P(A) \neq \Sigma_{k+1}^P(A)$.

We observe that in the above proof, the collapsing requirement $\Sigma_k^P(A) = BP\Sigma_k^P(A)$ can be satisfied when we diagonalize for separating requirements $\Sigma_h^P(A) \neq \Sigma_{h+1}^P(A)$ for different h, even if $h < k$. By a careful dovetailing of requirements $\Sigma_h^P(A) \neq \Sigma_{h+1}^P(A)$ for all $h > 0$, together with a complete encoding for $\Sigma_k^P(A) = BP\Sigma_k^P(A)$ for all $k > 0$, we obtain the following result:

Theorem 6.10. There exists an oracle A such that for every $k > 0$, $\Sigma_k^P(A) = BP\Sigma_k^P(A) \neq \Sigma_{k+1}^P(A)$.

Corollary 6.11. There exists an oracle A such that $co\text{-}NP(A) \not\subseteq AM_2(A)$.

There remains an interesting question of whether the above encoding technique still works when the polynomial hierarchy collapses to the $(k+1)$st level $\Sigma_{k+1}^P(A)$. A straightforward way of combining the encoding technique here with the diagonalization technique of Section 4.1 does not seem to work.

7. Other Hierarchies

In this section, we discuss briefly two other separation results on hierarchies which use similar proof techniques.

7.1. Generalized AM Hierarchy

In Section 1, we defined generalized polynomial hierarchy $\Sigma^P_{f(n)}$ by $f(n)$ levels of alternating $\exists_\mathrm{p}$- and $\forall_\mathrm{p}$-quantifiers. Instead of $\exists_\mathrm{p}$- and $\forall_\mathrm{p}$-quantifiers, we can also define a generalized AM-hierarchy by alternating $\exists^+_\mathrm{p}$- and $\exists_\mathrm{p}$-quantifiers. However, it is not clear that the complexity classes $AM_{f(n)}$ defined this way are equivalent to the languages defined by the *$f(n)$-round AM-game* as a generalization of the Arthur-Merlin game of Babai [1985]. The main difference is that in the Arthur-Merlin game, it is required that the total accepting probability be either $\geq 3/4$ (when Merlin wins) or $\leq 1/4$ (when Merlin loses), while in the definition by $\exists^+_\mathrm{p}$- and $\exists_\mathrm{p}$-quantifiers, it is only required that each individual quantifier $\exists^+_\mathrm{p}$ has a fixed probability bound. For a constant function $f(n) = k$, this difference is not substantial as the repetition of the same probabilistic computation can reduce the error probability (see, for instance, Zachos [1982]). However, the price to be paid for this reduction of error probability is the increased message length exchanged between Arthur and Merlin (i.e., the length of variables quantified by $\exists^+_\mathrm{p}$- and $\exists_\mathrm{p}$- quantifiers). When $f(n)$ is, for example, a linear function, the total length increase becomes intolerable (cf. Aiello, Goldwasser and Hastad [1986]).

Therefore, we define the generalized AM-hierarchy as follows:

$$
\begin{aligned}
AM_{f(n)}(A) = \{L \mid\ & (\exists P^1\text{-predicate } \sigma)(\forall x) \\
& [x \in L \Rightarrow (\exists^{++}_\mathrm{p} y_1)(\exists_\mathrm{p} y_2)\cdots(Q'_{f(|x|)} y_{f(|x|)})\ \sigma(A;\langle x, y_1, \cdots, y_{f(|x|)}\rangle)] \text{ and} \\
& [x \notin L \Rightarrow (\exists^{++}_\mathrm{p} y_1)(\forall_\mathrm{p} y_2)\cdots(Q''_{f(|x|)} y_{f(|x|)})\ \sigma(A;\langle x, y_1, \cdots, y_{f(|x|)}\rangle)]\},
\end{aligned}
$$

where $\exists^{++}_\mathrm{p} y$ denotes "for more than $(1-2^{-n})2^{q(n)}$ many $y \in \{0,1\}^{q(n)}$", $Q'_m = \exists^{++}_\mathrm{p}$ if m is odd, and $= \exists_\mathrm{p}$ if m is even, and $Q''_m = \exists^{++}_\mathrm{p}$ if m is odd and $= \forall_\mathrm{p}$ if m is even. It is easy to see that for all functions $f(n)$ which are bounded by polynomial functions, $AM_{f(n)}(A) \subseteq \Sigma^P_{f(n)}(A)$ for all sets A. On the other hand, the exact relation between the generalized AM-hierarchy and the polynomial hierarchy and the generalized polynomial hierarchy is not known. What we do know is that there exist oracles relative to which (a) the generalized AM-hierarchy does not contain the polynomial hierarchy (in fact, it does not even contain the class *co-NP*) and (b) each class $AM_{f(n)}$ in the generalized AM-hierarchy is not contained in $\Sigma^P_{g(n)}$ if $g(n) = o(f(n))$. In this section, we give brief outline of these proofs.

Theorem 7.1. Let f and g be two functions such that both are bounded by some polynomial function $q(n)$, and $g(n) = o(f(n))$. Then, there exists an oracle A such that $co\text{-}NP(A) \not\subseteq AM_{f(n)}(A) \not\subseteq \Sigma_{g(n)}(A)$.

Corollary 7.2. Let f_i be an infinite sequence of functions such that $f_i(n) = o(f_{i+1}(n))$ for all i and that each f_i is bounded by a polynomial function. Then, there exists an oracle A such that the classes $\Sigma^P_{f_i(n)}(A)$ form a proper infinite hierarchy between polynomial hierarchy $PH(A)$ and the class $PSPACE(A)$.

For the first part of Theorem 7.1, we need to show that there exists a set A such that $co\text{-}NP(A) \not\subseteq AM_{f(n)}(A)$. We let $L_A = \{0^n | (\forall y, |y| = n)\ 0^n y \in A\} \in co\text{-}NP(A)$, and show that L_A is not in any class $AM_{f(n)}(A)$.

First, following the approach of Section 2, we describe circuits corresponding to complexity classes $AM_{f(n)}(A)$. We define a new type of gates called MAJ$^+$ gates which operate as follows: a MAJ$^+$ gate outputs 1 (or 0) if more than $(1 - 2^{-n})\%$ of its inputs have value 1 (or 0, respectively), and it outputs ? otherwise. Then, define a $AM_{f(n)}(m)$-*circuit* to be a circuit of depth $f(n) + 2$ having the following structure: the top $f(n)$ levels of the circuit are alternating MAJ$^+$ and OR gates beginning with a top MAJ$^+$ gate, and the bottom two levels are OR of ANDs, and it has fanins $\leq 2^m$ and bottom fanin $\leq m$.

The proof of the following lemma is similar to that of Lemma 2.3. Call predicates (τ_1, τ_2) a pair of $AM_{f(n)}$-predicates if

$$\tau_1(A; x) \equiv (\exists^{++}_{\mathrm{p}} y_1)(\exists_{\mathrm{p}} y_2) \cdots (Q'_{f(n)} y_{f(n)}) \sigma(A; \langle x, y_1, \cdots, y_{f(n)} \rangle),$$

and

$$\tau_2(A; x) \equiv (\exists^{++}_{\mathrm{p}} y_1)(\forall_{\mathrm{p}} y_2) \cdots (Q''_{f(n)} y_{f(n)}) \sigma(A; \langle x, y_1, \cdots, y_{f(n)} \rangle),$$

for some P^1-predicate σ, where $n = |x|$, $Q'_m = \exists^{++}_{\mathrm{p}}$ if m is odd and $Q'_m = \exists_{\mathrm{p}}$ if m is even, and $Q'_m = \exists^{++}_{\mathrm{p}}$ if m is odd and $Q'_m = \forall_{\mathrm{p}}$ if m is even. Note that a pair of $AM_{f(n)}$-predicates (τ_1, τ_2) define a set in $AM_{f(n)}(A)$.

Lemma 7.3. Let $f(n)$ be a polynomially bounded function. For every pair of $AM_{f(n)}$-predicates (τ_1, τ_2) there is a polynomial q such that for every x, there exists a $AM_{f(n)}(q(|x|))$-circuit C, having the property that for any set A, $C\lceil_{\rho_A} = 1$ if $\tau_1(A; x)$ holds, and $C\lceil_{\rho_A} = 0$ if $\tau_2(A; x)$.

Following the diagonalization setting of Section 3, we see that the critical part of the proof is to show that for any set $B \in AM_{f(n)}(A)$, there exists a sufficiently large integer m such that $0^m \in L_A$ iff $0^m \notin B$. In other words, the following lemma on circuits suffices.

Lemma 7.4. Let f and q be two polynomial functions. Let C_0 be a depth-1 circuit which is the AND of $2^{n/2}$ many variables, and let C be a $AM_{f(n)}(q(n))$-circuit. Then, for sufficiently large n, there exists an assignment ρ such that $C\lceil_\rho \neq C_0\lceil_\rho$.

Sketch of proof. The proof is similar to the proof of Lemma 6.6. Let $v_1, \cdots, v_{2^{n/2}}$ be the variables of circuit C_0. Define assignments ρ_i, $0 \leq i \leq 2^{n/2}$, as follows: $\rho_0(v_j) = 1$ for all j, and for each $i \geq 1$, let $\rho_i(v_j) = 0$ iff $i = j$. We prove by induction on $m = f(n)$ that there must exist an i, $0 \leq i \leq 2^{n/2}$, such that $C\lceil_{\rho_i} \neq C_0\lceil_{\rho_i}$.

First, we consider the case when C has only two levels; i.e., C is the OR of $\leq 2^{q(n)}$ many ANDs, each having $\leq q(n)$ many inputs. Assume, by way of contradiction, that $C\lceil_{\rho_i} = C_0\lceil_{\rho_i}$ for all $i = 0, \cdots, 2^{n/2}$. Then, $C_0\lceil_{\rho_0} = 1$ implies $C\lceil_{\rho_0} = 1$ amd that in turn implies that there is an AND gate D of C having $D\lceil_{\rho_0} = 1$. However, D has only $\leq q(n)$ imputs and so, for sufficiently large n, there is at least one v_j such that neither v_j nor its negation $\overline{v}_j$ occurs in D. Therefore, $D\lceil_{\rho_j} = D\lceil_{\rho_0} = 1$. However, $D\lceil_{\rho_j} = 1$ implies $C\lceil_{\rho_j} = 1$ and this provides a contradiction.

Now, assume that C has $m > 2$ levels with the top gate being a MAJ$^+$ gate. By way of contradiction, suppose that $C\lceil_{\rho_i} = C_0\lceil_{\rho_i}$ for all i, $0 \leq i \leq 2^{n/2}$. Then, for each i, there are at most $(2^{-n})\%$ of the subcircuits D of C having $D\lceil_{\rho_i} \neq C_0\lceil_{\rho_i}$. Altogether, there are at most $(2^{-n} \cdot 2^{n/2})\%$ subcircuits computing differently from C_0 on at least one assignment ρ_i. That means that there exists at least one subcircuit D of C having $D\lceil_{\rho_i} = C_0\lceil_{\rho_i}$ for all i, $0 \leq i \leq 2^{n/2}$. Note that this subcircuit D has a top OR gate and that $D\lceil_{\rho_0} = 1$. Therefore, there exists at least one subcircuit G of D such that $G\lceil_{\rho_0} = 1$. Also, $G\lceil_{\rho_i} = 0$ for all $i \geq 1$, because $D\lceil_{\rho_i} = 0$ for all $i \geq 1$. So, we have shown that there is an $AM_{m-1}(q(n))$-circuit G such that $G\lceil_{\rho_i} = C_0\lceil_{\rho_i}$ for all i, $0 \leq i \leq 2^{n/2}$. This contradicts to the inductive hypothesis. □

For the second part of Theorem 7.1, we define $L_A = \{0^n | \ (\exists_{\mathrm{p}}^{++} y_1, |y_1| = n) (\exists_{\mathrm{p}} y_2, |y_2| = n) \cdots (Q'_{f(n)} y_{f(n)}, |y_{f(n)}| = n) \ [0^n y_1 \cdots y_{f(n)} \in A]\}$, where $Q'_m = \exists_{\mathrm{p}}^{++}$ if m is odd and $Q'_m = \exists_{\mathrm{p}}$ if m is even. From this definition, we do not know that $L_A \in AM_{f(n)}(A)$. What we need to do is to construct A to satisfy the additional condition that $x \notin L_A \iff \tau(A; x) \equiv (\exists_{\mathrm{p}}^{++} y_1)(\forall_{\mathrm{p}} y_2) \cdots (Q''_{f(n)} y_{f(n)}) \ 0^n y_1 \cdots y_{f(n)} \notin A$, where $Q''_m = \exists_{\mathrm{p}}^{++}$ if m is odd and $Q''_m = \forall_{\mathrm{p}}$ if m is even. In addition, we need to satisfy the requirements

$$R_i: (\exists x_i = 0^m)[0^m \in L_A \iff not\ \sigma_i(A; 0^m)],$$

where σ_i is the ith $\Sigma^{P,1}_{g(m)}$-predicate. For each m, define the following circuits: (a) C is the $\Sigma_{g(m)}(p_i(m))$-circuit corresponding to the predicate $\sigma_i(A; 0^m)$, and (b) C_0 is the $AM_{f(m)}$-circuit corresponding to the pair of predicates "$0^m \in L_A$" and $\tau(A; 0^m)$. Following the general diagonalization setting of Section 3, the separation problem is reduced to the following theorem on circuits.

Theorem 7.5. Let C and C_0 be defined as above. For sufficiently large m, there exists an assignment ρ such that $C\lceil_\rho \neq C_0\lceil_\rho \neq ?$.

The proof of the theorem again uses the technique of applying random restrictions ρ to circuits C and C_0 to simplify them. Two new issues arise in this application. First, a random restriction ρ is not likely to shrink circuit C_0 by only one level (like we had in Section 4.2), because we do not allow the MAJ$^+$ gates in C_0 to output ?. This is resolved by allowing ρ to shrink C_0 by k levels, where k is a constant depending on the degree of the polynomial $p_i(m)$ that bounds the bottom fanins of circuit C. Therefore, after applying random restrictions to these circuits for $g(m)$ times, circuit C is simplified into a simple depth-2 circuit but circuit C_0 still has $f(m) - k \cdot g(m)$ levels to perform diagonalization.

The second issue is how to define a probability space R of these restriction ρ such that (a) the Switching Lemma still holds with respect to space R, and (b) with a high probability, $C_0\lceil_\rho$ has depth at least $f(m) - k$. Since the circuit C_0 contains the MAJ$^+$ gates and since it is allowed to be shrunk by k levels, the probability space R is necessarily very complicated. Due to the space limit, we will omit the definition of the space R, as well as how it can satisfy the above two conditions. The interested reader should read Aiello, Goldwasser and Hastad [1986] for details.

7.2. The Low Hierarchy in NP

Schöning [1983] defined the high and low hierarchies within NP. It is natural to generalize it to the following relativized hierarchies. For $k \geq 0$, define

$$H^P_k(A) = \{L \in NP(A) \mid \Sigma^P_k(L \oplus A) = \Sigma^P_{k+1}(A)\},$$

$$L^P_k(A) = \{L \in NP(A) \mid \Sigma^P_k(L \oplus A) = \Sigma^P_k(A)\},$$

where $\oplus$ is the join operator on sets: $B \oplus C = \{0x \mid x \in B\} \cup \{1y \mid y \in C\}$. Let $HH(A) = \cup_{k\geq 0} H^P_k(A)$ and $LH(A) = \cup_{k \geq 0} L^P_k(A)$. It is not hard to see that

$H_k^P(A) \subseteq H_{k+1}^P(A)$ and $L_k^P(A) \subseteq L_{k+1}^P(A)$ for all $k \geq 0$ and for all A. It is however not known whether these hierarchies collapse or intersect each other. What we do know is that for all $k \geq 0$, $H_k^P(A) \cap L_k^P(A) \neq \emptyset$ iff $\Sigma_k^P(A) = \Pi_k^P(A)$ iff $NP(A) = H_k^P(A) = L_k^P(A)$. Other known relations about the hierarchies include: $L_0^P(A) = P(A)$, $L_1^P(A) = NP(A) \cup co\text{-}NP(A)$, $H_0^P(A) = \{L \mid L \text{ is } \leq_T^P\text{-complete for } NP(A)\}$. The interested reader is referred to Schöning [1983] and Ko and Schöning [1985] for more information about these hierarchies.

In this section, we show how to construct an oracle A such that $L_k^P(A) \neq L_{k+1}^P(A)$ for all $k \geq 0$. Similar separation result holds for the high hierarchy $HH(A)$. It is also possible to collapse both hierarchies to the kth level for any fixed $k \geq 0$. All these results are proven in Ko [1988b]. We only sketch the proof for the separation of the low hierarchy $LH(A)$.

In order to separate $L_{k+1}^P(A)$ from $L_k^P(A)$ for all $k \geq 1$, we need to find, for each $k \geq 0$, a set $B_k \in NP(A)$ satisfying the following two conditions:

(a_k) $\Sigma_{k+1}^P(B_k \oplus A) = \Sigma_{k+1}^P(A)$ and

(b_k) $\Sigma_k^P(B_k \oplus A) \neq \Sigma_k^P(A)$.

Note that each B_k is in $NP(A)$ and hence $\Sigma_k^P(B_k \oplus A) \subseteq \Sigma_{k+1}^P(A)$ and $\Sigma_{k+1}^P(B_k \oplus A) \subseteq \Sigma_{k+2}^P(A)$. This suggests that for condition (b_k) we need to separate the $(k+1)$st level of the polynomial hierarchy from the kth level, and for condition (a_k) we need to *partially* collapse the $(k+2)$nd level of the polynomial hierarchy to the $(k+1)$st level. The collapsing part is only a partial collapsing because condition (b_{k+1}) implies the separation of the $(k+2)$nd level of the polynomial hierarchy from the $(k+1)$st level. It is interesting to point out that our goal is only a separation result but our proof technique is more like the one used in Section 6.2 for collapsing results.

How do we *partially* collapse the polynomial hierarchy? This involves a careful choice of diagonalization regions and the witness sets B_k. To satisfy condition (a_k), we would like to make the set B_k to behave like an empty set as much as possible, and, on the other hand, to satisfy condition (b_k), we need to make the set B_k to be similar to, for instance, the set L_A used in Section 3. More precisely, we define it as follows. First, let $e(0) = 1$, and $e(n+1) = 2^{2^{e(n)}}$ for all $n > 0$. Then, for each $k \geq 0$, let

$$B_k = \{x \mid |x| = e(\langle k, m \rangle) \text{ for some } m, \text{ and } (\exists y, |y| = |x|)\ 0xy \in A\}.$$

That is, we predetermine the diagonalization regions for all diagonalization processes for condition (b_k) for all k (namely, for any k, the corresponding regions locate close to $e(\langle k,m\rangle)$ for some m), and make B_k to be identical to the empty set outside these regions.

This definition achieves two important subgoals for our construction. First, it separates the diagonalization regions for conditions (b_k) from that for conditions (b_h) if $k \neq h$. Second, it satisfies condition (a_k) immediately outside the diagonalization regions for (b_k). This allows us to divide and conquer the numerous seemingly contradictory requirements.

Now we state our requirements as follows. For the separation part, we need

$R_{k,n,0}$: $(\exists x_n)[x_n \in E_A^{(k)} \iff \sigma_n^{(k)}(A; x_n) \text{ is false}]$,

for some set $E_A^{(k)} \in \Sigma_k^P(B_k \oplus A)$, where $\sigma_n^{(k)}$ is the nth $\Sigma_k^{P,1}$-predicate. We simply let $E_A^{(k)} = \{0^{e(\langle k,m\rangle)} \mid (\exists y_1, |y_1| = r)\cdots(Q_k y_k, |y_k| = r)\ y_1 \cdots y_k 0^t \in B_k, 0 \le t < k, rk + t = e(\langle k,m\rangle)\}$.

For the collapsing part, we need

$R_{k,n,1}$: $(\forall x, |x| = n)[x \in K^{k+1}(B_k \oplus A) \iff \tau_k(A; x)]$,

for some $\Sigma_{k+1}^{P,1}$-predicate τ_k. (Recall that $K^n(A)$ is a standard $\Sigma_k^P(A)$-complete set.)

Note that for each x such that $2^{e(\langle k,m\rangle)} \le |x| < e(\langle k, m+1\rangle)$ for some m, the question of whether $x \in K^{k+1}(B_k \oplus A)$ can be simulated by a Σ_{k+1}^P-machine using only A as the oracle: It simulates the computation of $x \in K^{k+1}(B_k \oplus A)$ and answer each query "$y \in ? B_k$" as follows: if $|y| \neq e(\langle k,r\rangle)$ for any $r \le m$, then answer NO, else answer YES iff $(\exists z, |z| = |y|)0yz \in A$. So, requirements $R_{k,n,1}$ need to be satisfied only if $e(\langle k,m\rangle) \le n < 2^{e(\langle k,m\rangle)}$. We let $\tau_k(A; x)$ be the following predicate:

$$\tau_k(A; x) \equiv (\exists u_1, |u_1| = |x|)\cdots(Q_{k+1} u_{k+1}, |u_{k+1}| = |x|)\ 10^k 1 x u_1 \cdots u_{k+1} \in A.$$

(The heading of 10^k1 is used to distinguish between τ_k and τ_h when $h \neq k$.)

Then, in stage $n = \langle k,m\rangle$, we find the least i such that requirement $R_{k,i,0}$ is not yet satisfied and try to satisfy it by witness $x_i = 0^{e(n)}$ and the diagonalization region $D_k(i) = \{y \mid e(n) \le |y| < 2^{e(n)}\}$. In the meantime, we need to satisfy requirements $R_{h,j,1}$ for all $h \ge 0$ and all j, $e(n) \le j < 2^{e(n)}$. Observe that if $h \neq k$ then, $2^{e(\langle h,t\rangle)} \le j < e(\langle h, t+1\rangle)$ for some t if $e(n) \le j < 2^{e(n)}$. Therefore, we need only to worry about requirements $R_{k,j,1}$ for $e(n) \le j < 2^{e(n)}$.

It is easy to check now that all these preparations lead us to a familiar diagonalization setup of Section 6.1: diagonalizing against a $\Sigma_k^{P,1}$-predicate while keeping some set in $\Sigma_{k+2}^P(A)$ encoded by a $\Sigma_{k+1}^{P,1}$-predicate. The separation of $LH(A)$ follows immediately.

Theorem 7.6. There exists an oracle A such that for all $k \geq 0$ $L_{k+1}^P(A) \neq L_k^P(A)$.

8. Conclusion

In this paper, we have presented a general method of separating or collapsing hierarchies by oracles. The construction of the oracles usually involves two different types of proof techniques: the recursion-theoretic one and the combinatorial one. The simple relations between the computation trees generated by oracle machines and the circuits with unbounded fanins provide nice reduction of recursion-theoretic problems to combinatorial problems. For the pure separation results, the recursion-theoretic part is usually a simple diagonalization, and the main difficulty arises from finding good lower bounds for circuit complexity. For the collapsing results, both the recursion-theoretic setup and the combinatorial techniques become more complicated. It is often necessary, like in Sections 6.2 and 6.3, to use more ad hoc tricks to obtain the required lower bound results.

Following this point of view, we may continue this research in two directions. First, we need to make a deeper investigation into the diagonalization and encoding techniques, particularly how two techniques can be combined to satisfy more seemingly contradictory requirements. Do more powerful recursion-theoretic techniques, such as the finite-injury method, possibly have interesting applications in this type of proofs? Are there better forms of encoding of information to provide more free space in diagonalization regions? Although the diagonalization and encoding techniques have been examined by many people, many ad hoc techniques still seem beyond our understanding. (One example is the question posed in Section 6.3: to construct an oracle A such that $BPH(A)$ collapses to $PH(A)$ and $PH(A)$ collapses to the kth level.)

Second, in the combinatorial side, we would like to see how far the proof technique of Yao and Hastad can be stretched. Can we find general conditions on the probability spaces which allow the Switching Lemma to hold? Are there totally different approaches (such as the one of Smolensky [1987]) that make lower bound proofs easier or give better lower bounds? Even more, what is the limit of this types of combinatorial arguments? For instance, the question of whether the polynomial hierarchy is infinite relative to a random oracle is still open. Can we sharpen this proof technique to solve this problem? or do we really need new ideas?

9. References

9.1. Bibliographic Notes

Section 1. Hopcroft and Ullman [1979] and Garey and Johnson [1979] contain introductory materials on complexity classes *P, NP, PSPACE* and *PH*. They also include formal models for oracle machines. A more recent and more complete textbook on complexity classes is Balcázar, Diaz and Gabarró [1988], which also contains the formal definitions of probabilistic classes *R* and *BPP*. The *AM* hierarchy was introduced by Babai [1985] and the interactive proof systems by Goldwasser, Micali and Rackoff [1985]. Their equivalence was proved by Goldwasser and Sipser [1986]. The *BP* operator, the probabilistic polynomial hierarchy and its relation to the class *AM* were given by Schöning [1987]. The nice layout of Figure 1 is from Tang and Watanabe [1988]. The notation $\exists_{\mathrm{p}}^{+}$ is due to Zachos [1986], which contains a survey on the relations between complexity classes definable by the $\exists_{\mathrm{p}}^{+}$-quantifiers over polynomial-time predicate. The relativization of these complexity classes is most often done by adding oracles to corresponding machine models; e.g., class $\Sigma_k^P(A)$ is defined by alternating machines with oracles which can make at most k alternations [Chandra, Kozen and Stockmeyer, 1981]. Our approach essentially cleans up the computation of those oracle machines and pushes the queries down to the bottom level (cf. Furst, Saxe and Sipser [1984]).

Section 2. The idea of using circuits to represent oracle computation trees (Lemma 2.3) is originated from Furst, Saxe and Sipser [1984]. They also introduced the concept of random restrictions and gave the first super-polynomial (but still sub-exponential) lower bound for constant-depth parity circuits. Sipser [1983] pointed out that the relation given by Lemma 2.3 may be applied to the separation of $PH(A)$. Majority gate MAJ, as well as similar threshold gates, have been considered in circuit complexity theory. See, for example, Hajnai et al [1987].

Section 3. The first application of the diagonalization technique to relativization was by Baker, Gill and Solovay [1975]. Many people discussed this application, including Angluin [1980], Bennett and Gill [1981], Kozen [1978], and Torenvliet and van Amde Boas [1986].

Section 4. Before the breakthrough of Yao [1985], $PH(A)$ has been known to extend to at least $\Sigma_3^P(A)$. These results are due to Baker, Gill and Solovay [1975], Baker and Selman [1979] and Heller [1984]. Angluin [1980] also showed that $P(\#P(A)) \not\subseteq \Sigma_2^P(A)$ (a special case of Corollary 4.5). The first exponential lower bound for constant-depth parity circuit was proved by Yao [1985]. He also claimed, without a proof, a similar exponential lower bound on depth-k circuit for function f_{k+1}^n. Hastad [1986, 1987] gave a simpler proof for parity circuit, and achieved the almost optimal bound of Theorem 4.1. He also proved Yao's claim using the same proof technique (Theorem 4.6). Smolensky [1987] used an algebraic method to give a much shorter proof for the exponential lower bound for parity circuit, but his method does not seem to work for f_{k+1}^n functions. More recently, Du [1988] found, based on Hastad's proofs, simpler proofs of the Switching Lemmas (Lemmas 4.2 and 4.7), which also yield a slightly better lower bound for parity circuits. Our proofs in Sections 4.1 and 4.2 are based on Hastad's proofs. The class $\oplus P(A)$ is first defined in Papadimitriou and Zachos [1983]. Recently, Toran [1988] has proved that $NP(A) \not\subseteq \oplus P(A)$ relative to some oracle A. Lemma 4.10 was first proved by Baker and Selman [1979] in a different form. Theorem 4.11 was proved in Ko [1988a]. Corollary 4.12 has been observed independently by Watanabe [1987].

Section 5. Our example is one of the first application of the encoding technique to relativization appeared in Baker, Gill and Solovay [1975].

Section 6. The main results in Sections 6.1 and 6.2 are from Ko [1989] and the ones in Section 6.3 are from Ko [1988a]. Weaker collapsing results before include Baker, Gill and Solovay [1975] (collapsing $PSPACE$ to P, and collapsing Σ_2^P to NP), Rackoff [1982] and Bennett and Gill [1981] (collapsing R to P but keep $P \neq NP$). The combinatorial lemma Lemma 6.7 has been known to many researchers, including Adleman [1978], Bennett and Gill [1981], Ko [1982] and Zachos [1982].

Section 7. The AM hierarchy was introduced by Babai [1985], who showed that the bounded AM hierarchy collapses to AM_2 and conjectured that the generalized one also collapses to AM_2. The use of notation $\exists_{\mathrm{p}}^{++}$ and the MAJ^+ gate is new. First part of Theorem 7.1 is due to Fortnow and Sipser [1988] and the second part due to Aiello, Goldwasser and Hastad [1986]. The high and low hierarchies in NP was introduced by Schöning [1983], who proved some basic properties of these hierarchies. Ko and Schöning [1985] contains more classification of low sets by structural properties. The proofs in Section 7.2 are from Ko [1988b].

Section 8. Cai [1986] and Babai [1987] proved that *PSPACE* is not in *PH* relative to a random oracle. Other separation results by random oracles are in Bennett and Gill [1981]. Babai [1985] and Hastad [1987] have pointed out the difficulty of using Hastad's technique to prove that *PH* is infinite relative to a random oracle.

9.2. Bibliography

Adleman, L. [1978], Two theorems on random polynomial time, *Proc. 19th IEEE Symp. on Foundations of Computer Science*, 75–83.

Aiello, W., Goldwasser, S. and Hastad, J. [1986], On the power of interaction, *Proc. 27th IEEE Symp. on Foundations of Computer Science*, 368–379.

Angluin, D. [1980], On counting problems and the polynomial-time hierarchy, *Theoret. Comput. Sci.* **12**, 161–173.

Babai, L. [1985], Trading group theory for randomness, *Proc. 17th ACM Symp. on Theory of Computing*, 421–429.

Babai, L. [1987], Random oracle separates PSPACE from the polynomial-time hierarchy, *Inform. Process. Lett.* **26**, 51–53.

Baker, T., Gill, J. and Solovay, R. [1975], Relativizations of the *P=?NP* question, *SIAM J. Comput.* **4**, 431–442.

Baker, T. and Selman, A. [1979], A second step toward the polynomial hierarchy, *Theoret. Comput. Sci.* **8**, 177–187.

Balcázar, J. [1985], Simplicity, relativizations, and nondeterminism, *SIAM J. Comput.* **14**, 148–157.

Balcázar, J., Book, R. and Schöning, U. [1986], The polynomial-time hierarchy and sparse oracles, *J. Assoc. Comput. Mach.* **33**, 603–617.

Balcázar, J., Diaz, J. and Gabarró, J. [1988], *Structural Complexity I*, Springer-Verlag, Berlin.

Balcázar, J. and Russo, D. [1988], Immunity and simplicity in relativizations of probabilistic complexity classes, *RAIRO Theoret. Inform. and Appl.* **22**, 227–244.

Bennett, C. and Gill, J. [1981], Relative to a random oracle, $P^A \neq NP^A \neq coNP^A$ with probability 1, *SIAM J. Comput.* **10**, 96–113.

Boppana, R., Hastad, J. and Žachos, S. [1987], Does co-NP have short interactive proofs?, *Inform. Process. Lett.* **25**, 127–132.

Buss, J. [1986], Relativized alternation, *Proc. Structure in Complexity Theory Conf.*, Lecture Notes in Computer Science, **223**, Springer, 66–76.

Cai, J. [1986], With probability one, a random oracle separates PSPACE from the polynomial-time hierarchy, *Proc. 18th ACM Symp. on Theory of Computing*, 21–29.

Chandra, A., Kozen, D. and Stockmeyer, L. [1981], Alternation, *J. Assoc. Comput. Mach.* **28**, 114–133.

Du, D. [1988], personal communication.

Fortnow, L. and Sipser, M. [1988], Are there interactive protocols for *co-NP* languages?, *Inform. Process. Lett.* **28**, 249–252.

Furst, M., Saxe, J. and Sipser, M. [1984], Parity, circuits, and the polynomial time hierarchy, *Math. Systems Theory* **17**, 13–27.

Garey, M. and Johnson, D. [1979], *Computers and Intractability, a Guide to the Theory of NP-Completeness*, Freeman, San Francisco.

Goldwasser, S., Micali, S. and Rackoff, C. [1985], The knowledge complexity of interactive proof systems, *Proc. 17th ACM Symp. on Theory of Computing*, 291–304.

Goldwasser, S. and Sipser, M. [1986], Private coins versus public coins in interactive proof systems, *Proc. 18th ACM Symp. on Theory of Computing*, 59–68.

Hajnai, A., Maass, W., Pudlak, P., Szegedy, M. and Turan, G. [1987], Threshold circuits of bounded depth, *Proc. 28th IEEE Symp. on Foundations of Computer Science*, 99–110.

Hastad, J. [1986], Almost optimal lower bounds for small depth circuits, *Proc. of 18th ACM Symp. on Theory of Computing*, 6–20.

Hastad, J. [1987], *Computational Limitations for Small-Depth Circuits*, (Ph.D. Dissertation, MIT), MIT Press, Cambridge.

Heller, H. [1984], Relativized polynomial hierarchies extending two levels, *Math. Systems Theory* **17**, 71–84.

Hopcroft, J. and Ullman, J. [1979], *Introduction to Automata Theory, Languages, and Computation*, Addison-Wesley, Reading.

Ko, K. [1982], Some observations on probabilistic algorithms and NP-hard problems, *Inform. Process. Lett.* **14**, 39–43.

Ko, K. [1988a], Separating and collapsing results on the relativized probabilistic polynomial time hierarchy, preprint.

Ko, K. [1988b], Separating the low and high hierarchies by oracles, preprint.

Ko, K. [1989], Relativized polynomial time hierarchies having exactly k levels, *SIAM J. Comput.*, in press; also in *Proc. 20th ACM Symposium on Theory of Computing* [1988], 245–253.

Ko, K. and Schöning, U. [1985], On circuit-size complexity and the low hierarchy in *NP*, *SIAM J. Comput.* **14**, 41–51.

Kozen, D. [1978], Indexing of subrecursive classes, *Proc. 10th ACM symp. on Theory of Computing*, 89–97.

Long, T. and Selman, A. [1986], Relativizing complexity classes with sparse oracles, *J. Assoc. Comput. Mach.* **33**, 618-627.

Papadimitriou, C. and Zachos, S. [1983], Two remarks on the power of counting, *Proc. 6th GI Conf. on Theoretical Computer Science*, Lecture Notes in Computer Science **145**, 269–276.

Rackoff, C. [1982], Relativized questions involving probabilistic algorithms, *J. Assoc. Comput. Mach.* **29**, 261–268.

Schöning, U. [1983], A low and a high hierarchy within *NP*, *J. Comput. System Sci.* **27**, 14–28.

Schöning, U. [1987], Probabilistic complexity classes and lowness, *Proc. 2nd IEEE Structure in Complexity Theory Conf.*, 2–8.

Sipser, M. [1983], Borel sets and circuit complexity, *Proc. of 15th ACM Symp. on Theory of Computing*, 61–69.

Smolensky, R. [1987], Algebraic methods in the theory of lower bounds for boolean circuit complexity, *Proc. 19th ACM Symp. on Theory of Computing*, 77–82.

Torán, J. [1988], *Structural Properties of the Counting Hierarchies*, Doctoral Dissertation, Facultat d'Informatica, UPC Barcelona.

Tang, S. and Watanabe, O. [1988], On tally relativizations of BP-complexity classes, *Proc. 3rd IEEE Structure in Complexity Theory Conf.*, 10–18.

Torenvliet, L. and van Emde Boas, P. [1986], Diagonalization methods in a polynomial setting, *Proc. Structure in Complexity Theory Conf.*, Lecture Notes in Computer Science **223**, 330–346.

Watanabe, O. [1987], Personal communication.

Wilson, C. [1988], A measure of relativized space which is faithful with respect to depth, *J. Comput. System Sci.* **36**, 303–312.

Yao, A. [1985], Separating the polynomial-time hierarchy by oracles, *Proc. of 26th IEEE Symp. on Foundations of Computer Science*, 1–10.

Zachos, S. [1982], Robustness of probabilistic computational complexity classes under definitional perturbations, *Inform. Contr.* **54**, 143–154.

Zachos, S. [1986], Probabilistic quantifiers, adversaries, and complexity classes, *Proc. of Structure in Complexity Theory Conf.*, 383–400.

RANDOMNESS, TALLY SETS, AND COMPLEXITY CLASSES[1]

TANG Shouwen

Beijing Computer Institute
Beijing 100044
People's Republic of China

I. Introduction.

Baker, Gill, and Solovay [6] proved that there is an oracle A such that P(A) = NP(A) and there is an oracle B such that P(B) ≠ NP(B). It is natural to ask how many oracles separate P and NP, and how many oracles make them equal. Another question arises: how can one count the number of oracles (languages) with a certain property? Bennett and Gill [8] introduced the concept of random oracle (random language). Because of this concept, we can say that "almost every language has property ..." if, with probability one, a random oracle has this property. Bennett and Gill established the following results.

Proposition 1.1. For almost every oracle A, P(A) ≠ NP(A).

Proposition 1.2. For almost every oracle A, P(A) = BPP(A).

Furthermore, Bennett and Gill proposed a "random oracle hypothesis" (ROH), which we describe informally here. ROH asserts that a relation for complexity classes holds in the unrelativized world if and only if with probability one it holds for random oracles in the relativized world.

---

[1] This paper is based on a lecture given at the International Symposium on Combinatorial Optimization, held at the Nankai Institute of Mathematics, Tianjin, People's Republic of China, August 1988.

As an example, consider the $P =? NP$ problem. In order to find an oracle B separating P and NP, B should be chosen carefully to guarantee that B is biased to the desired separation. The same is true for finding an A such that $P(A) = NP(A)$. If the oracle is chosen randomly, then it is not biased to any side, i.e., it does not help in the separation of P and NP. Since a random oracle separates P and NP (Proposition 1.1) and a random oracle is not biased, it should be the case that $P \neq NP$.

Unfortunately, most computer scientists refute ROH because of a series of counterexamples [13, 17, 21]. Some of these counterexamples can be described in the following way.

Proposition 1.3. APSPACE = EXP, but for almost every oracle A, APSPACE(A) $\neq$ EXP(A), where APSPACE is the class of languages accepted by alternating polynomial space Turing machines and EXP is the class of languages accepted by deterministic Turing machines in time that is exponential in a polynomial.

Proposition 1.4. PQUERY = PSPACE, but for almost every oracle A, PQUERY(A) $\neq$ PSPACE(A), where PQUERY(A) is the class of languages accepted by polynomial space oracle machines which make only polynomial many queries and PQUERY = PQUERY($\emptyset$).

While studying relations between relativizations of complexity classes, Cai [11] and Babai [5] established the following result.

Proposition 1.5. For almost every oracle A, PH(A) $\neq$ PSPACE(A).

Another direction of research about random oracles concerns the characterization of membership in certain complexity classes. For example, we have the following:

Proposition 1.6. [1, 8, 17] $A \in$ BPP if and only if for almost every oracle B, $A \in P(B)$.

Proposition 1.7. [1] $A \in P$ if and only if for almost every oracle B, $A \leq_m^P B$.

In addition to the characterizations of membership in P and BPP, these two propositions tell us the difference between polynomial time many-one and Turing reducibilities with respect to a random oracle.

On the other hand previous studies show that relativization with respect to a tally set (i.e., a set over a one letter alphabet) is much closer to the unrelativized case than that with respect to a general oracle (i.e., a set over a two letter alphabet). As examples, there are the following facts.

Proposition 1.8. [19] $P \neq NP$ if and only if there is a tally set T such that $P(T) \neq NP(T)$.

Proposition 1.9. [19] The following are equivalent:

a. the polynomial-time hierarchy collapses;
b. there is a tally set T such that the polynomial-time hierarchy relative T collapses;
c. for every tally set T the polynomial-time hierarchy relative to T collapses.

Proposition 1.8 implies that although we know that almost every oracle separates P and NP, the existence of a single tally oracle with this property is not known. From this point of view, tally oracles should be studied in detail in order to obtain some insight into this fundamental problem. Indeed, sets that are Turing reducible or Turing equivalent or polynomially isomorphic to tally sets play important roles in complexity theory [2, 3, 9, 10, 25].

In this paper we shall not discuss the ROH. Instead, we describe results on randomness and tally sets by defining the concept of random tally sets and then summarizing some applications of randomness to the study of complexity classes.

2. Preliminaries.

Let $\Sigma = \{0, 1\}^*$ so that Σ^* is the set of all finite strings over Σ. A word is an element of Σ^*, and a language is a subset of Σ^*. For a word x, $|x|$ is the length of x. A <u>tally</u> language is a subset of $\{0\}^*$. The class of all tally languages is denoted by TALLY: TALLY = POWER($\{0\}^*$). The set of all one-way infinite sequences over Σ is denoted by Ω.

For a class **CL** of languages, co-**CL** = $\{A \mid (\Sigma^* - A) \in$ **CL**$\}$. For a class **TC** of tally languages, co-**TL** = $\{ T \mid (\{0\}^* - T) \in$ **TL**$\}$.

We fix some pairing function $\langle\, ,\, \rangle: \Sigma^* \times \Sigma^* \rightarrow \Sigma^*$ such that both the function and its inverse are computable in polynomial time.

For any language A, P(A) is the class of <u>languages accepted by deterministic polynomial time oracle machines relative to</u> A and NP(A) is the class of <u>languages accepted by nondeterministic polynomial time oracle machines relative to</u> A. For a class **CL** of languages, P(**CL**) = $\cup_{A \in \mathbf{CL}}$P(A) and NP(**CL**) = $\cup_{A \in \mathbf{CL}}$NP(A).

The <u>polynomial-time hierarchy</u> is defined as follows (see [24, 28]):

$$\Sigma_0^P = \Delta_0^P = \Delta_1^P = \Pi_0^P = P; \quad \Sigma_{n+1}^P = NP(\Sigma_n^P); \quad \Delta_{n+1}^P = P(\Sigma_n^P);$$

$$\Pi_n^P = \text{co-}\Sigma_n^P; \text{ and } PH = \cup_{n \geq 0} \Sigma_n^P.$$

PSPACE(A) is the class of <u>languages accepted by polynomial space-bounded oracle Turing machines relative to</u> A, where the query tape is counted in the space bound; clearly, PSPACE = PSPACE($\emptyset$).

The following facts are well known.

Proposition 2.1.

a. For $n \geq 0$, $\Sigma_n^P \subseteq \Delta_{n+1}^P \subseteq \Sigma_{n+1}^P \cap \Pi_{n+1}^P$.

b. For $n \geq 0$ and $m \geq 1$, $\Sigma_{n+m}^P = \Sigma_n^P(\Sigma_m^P)$.

c. For $n \geq 1$, $\Sigma_n^P \subseteq \Pi_n^P$ if and only if $\Pi_n^P \subseteq \Sigma_n^P$ if and only if $PH = \Sigma_n^P = \Pi_n^P$. In this case, we say that the polynomial-time hierarchy *collapses* .

d. For $n \geq 0$, $NP(\Sigma_n^P \cap \Pi_n^P) = \Sigma_n^P$.

e. $PH \subseteq PSPACE$, and if equality holds, then the polynomial-time hierarchy collapses.

Language A is said to have polynomial size circuits if there is a polynomial $p(n)$ such that for each n, there is an n-input circuit C_n of size at most $p(n)$ with the property that for any word x of length n, $C_n(x) = 1$ if and only if $x \in A$.

Let **CL** be a class of languages. We say that language A has polynomial size advice with respect to **CL** if there is a function h from the natural numbers to Σ^* and a set $D \in$ **CL** with the properties

(i) there is a fixed polynomial $p(n)$ such that for all n $|h(n)| \leq p(n)$, and

(ii) for every x, $x \in A$ if and only if $\langle x, h(|x|)\rangle \in D$.

The class of languages having polynomial size advice with respect to **CL** is denoted by **CL**/poly.

The following facts are well known.

Proposition 2.2. The following are equivalent:
a. A has polynomial size circuits;
b. A ε P/poly;
c. A ε P(TALLY).

Proposition 2.3. For **CL** = Σ_n^P, Π_n^P, Δ_n^P, or PSPACE,

A ε **CL**/poly if and only if A ε **CL**(TALLY).

For example, A ε NP/poly if and only if there is a tally language T such that A ε NP(T). This can be interpreted as saying that A has polynomial generators (or nondeterministic polynomial size circuits).

3. Random Tally Sets.

Recall that Ω denotes the set of all one-way infinite sequences over Σ. Thus, Ω is the infinite product of Σ.

For every tally set T, the characteristic sequence of T, $b_0b_1b_2b_3$ ... , is an element of Ω; recall that $b_i = 1$ if 0^i ε T and $b_i = 0$ otherwise. This provides a one-to-one mapping of TALLY onto Ω. We shall identify a tally set with its characteristic sequence. Therefore, a subset of Ω is identified with a class of tally sets.

In Σ, define $\mu(0) = \mu(1) = 1/2$ so that Σ is a discrete probability space. By taking the completion of the infinite product of this probability space, we obtain a measure or probability on Ω. We will again use μ for this measure. Since we identify a tally set with its characteristic sequence (an element of Ω), we can say that a class of tally sets is measurable (or not) and its measure or probability is defined as the measure of the corresponding subset of Ω.

There is a natural way to regard an element of Ω as a real number in the unit interval [0, 1]. It is not difficult to see that the above measure μ in Ω is just the Lebesgue measure in the unit interval.

Let $n \geq 0$ be any fixed integer. Since $\mu\{b_0b_1b_2b_3 \dots \mid b_n = 1\} = 1/2$, we see that $\mu\{T \mid 0^n \in T\} = 1/2$. This equality can be interpreted as the fact that with probability 1/2 a random tally set contains 0^n. Observing that this is true for all $n \geq 0$, we see the following: a random tally set is obtained by an independent series of tosses of an unbiased coin, i.e., the result of the nth coin toss determines whether the nth word of $\{0\}^*$ is put into the set. This coincides with our intuition for randomness: a random tally set is obtained randomly.

If a property (or a class) of tally sets has measure 1, then we say that almost every tally set has this property or that this property holds for almost every tally sets.

The following proposition is a special case of the 0-1 Law in probability theory. It is a useful tool in the study of random oracles. Although it is not used in this paper since we will not prove any theorem in the following, we think that it is fitting and proper to mention it here because of its importance.

We say that a class of tally sets is closed under finite variation if whenever a set A belongs to this class and B and A have a finite symmetric difference, then B belongs to the class.

Proposition 3.1. 0-1 Law: Every measurable class of tally sets which is closed under finite variation has measure either 0 or 1.

4. The BP( ) Operator.

Gill [12] studied the probabilistic Turing machine and the probabilistic complexity classes BPP, R, and ZPP. Babai [4] defined the Arthur-Merlin game and the complexity classes AM and AM(poly). We will use the following equivalent definitions for BPP and for AM.

Let S be a property of words and let m be a natural number. Define $Pr_m[y \mid y$ has property $S]$ to be $\|\{y \mid y$ has property S and $|y| = m\}\| /2^m$. Thus, $Pr_m[y \mid y$ has property $S]$ is the probability of S under the condition that the strings have length m.

BPP is the class defined as follows: $A \in BPP$ if and only if there is a set $B \in P$ and a polynomial $p(n)$ such that
if $x \in A$, then $Pr_{p(|x|)}[y \mid \langle x, y\rangle \in B] \geq 3/4$, and
if $x \notin A$, then $Pr_{p(|x|)}[y \mid \langle x, y\rangle \notin B] \geq 3/4$.

AM is the class defined as follows: $A \in AM$ if and only if there is a set $B \in NP$ and a polynomial $p(n)$ such that
if $x \in A$, then $Pr_{p(|x|)}[y \mid \langle x, y\rangle \in B] \geq 3/4$, and

if $x \notin A$, then $Pr_{p(|x|)}[y \mid \langle x, y\rangle \notin B] \geq 3/4$.

Schöning [22] generalized these concepts by defining the BP operator in the following way.

Let **CL** be a class of languages. Define BP·**CL** as follows: $A \in$ BP·**CL** if and only there is a set $B \in$ **CL** and a polynomial $p(n)$ such that
if $x \in A$, then $Pr_{p(|x|)}[y \mid \langle x, y\rangle \in B] \geq 3/4$, and

if $x \notin A$, then $Pr_{p(|x|)}[y \mid \langle x, y\rangle \notin B] \geq 3/4$.

It is not difficult to prove the following facts.

Proposition 4.1. For any classes CL, CL_1, CL_2,

a. BP·co-CL = co-(BP·CL);
b. $CL_1 \subseteq CL_2$ implies that BP·$CL_1 \subseteq$ BP·CL_2;
c. if CL is closed under padding (i.e., A ε CL implies that ⟨A, Σ*⟩ ε CL, where ⟨A, Σ*⟩ = {⟨x, ⟩ | x ε A, y ε Σ*}), then CL $\subseteq$ BP·CL.

Obviously, BPP = BP·P and AM = BP·NP. Now for each n, we have the classes BP·Σ_n^P, BP·Π_n^P, and BP·Δ_n^P. This collection of classes makes up the "probabilistic" polynomial-time hierarchy.

Proposition 4.2. BP·PH = PH and BP·PSPACE = PSPACE.

Using hashing techniques, Sipser [23] proved the following:

Proposition 4.3. BPP $\subseteq \Sigma_2^P$; hence, BPP $\subseteq \Sigma_2^P \cap \Pi_2^P$.

Schöning [22] proved the following:

Proposition 4.4. For $n \geq 1$, BP·$\Sigma_n^P \subseteq \Pi_{n+1}^P$ and BP·$\Pi_n^P \subseteq \Sigma_{n+1}^P$.

Some of these relationships are summarized in Figure 1.

Now we describe some evidence that for each n the classes Σ_n^P and BP·Σ_n^P are very close to each other.

Proposition 4.5. [27] Let **CL** be any of Σ_n^P, Π_n^P, Δ_n^P, PH, or PSPACE.

a. A ε BP·CL if and only if for almost every tally set T, A ε CL(T);
b. For almost every tally set T, CL(T) = BP·CL(T).

If **CL** is P or NP, then Proposition 4.5 becomes:

Proposition 4.6.

a. A ε BPP if and only if for almost every tally set T, A ε P(T).
b. A ε AM if and only if for almost every tally set T, A ε NP(T).
c. For almost every tally set T, P(T) = BPP(T).
d. For almost every tally set T, NP(T) = AM(T).

Statement 4.6(a) is the tally version of Proposition 1.2. Observe that, as noted after Proposition 1.9, the results about random tally oracle sets are not just special cases of the results about random (general) oracle sets.

Combining 4.6(a) with Proposition 2.2, we see that A ε BPP implies that A has polynomial size circuits, i.e., if for almost every tally set T, A ε P(T), then for some tally set T, A ε P(T). If we interpret the tally set T in Proposition 2.2(c) as the polynomial size circuits and the phrase "a lot of" as measure 1 (or probability 1), then the measure on tally sets gives us a formal way to "count" the number of polynomial size circuits. Proposition 4.6(a) asserts that A is in BPP if and only if A has a lot of polynomial size circuits. At first glance this is quite surprising since a set having polynomial size circuits is not necessarily recursive; there are countably many recursive sets but uncountable many sets having polynomial size circuits.

Proposition 4.6(a) can be interpreted as $A \in AM$ if and only if A has a lot of polynomial size generators (or nondeterministic polynomial size circuits).

These remarks lead to Propositions 4.7 and 4.8.

Proposition 4.7.

a. For $n \geq 0$ and $m \geq 1$, $BP\cdot \Sigma^P_{m+n} = BP\cdot \Sigma^P_m(BP\cdot \Sigma^P_n)$.

b. For $n \geq 1$, $BP\cdot\Sigma^P_n \subseteq BP\cdot\Pi^P_n$ if and only if $BP\cdot\Pi^P_n \subseteq BP\cdot\Sigma^P_n$ if and only if $PH = BP\cdot\Sigma^P_n = BP\cdot\Pi^P_n$.

c. For $n \geq 0$, $BP\cdot NP(BP\cdot\Sigma^P_n \cap BP\cdot\Pi^P_n) = BP\cdot\Sigma^P_n$.

This shows that the probabilistic polynomial-time hierarchy has properties that parallel that of the usual polynomial-time hierarchy. Statement 4.7(a) generalizes the result of Ko [16] that BPP(BPP) = BPP, and the result of Zachos and Furer [29] that AM(BPP) = AM.

The proof technique of Proposition 4.7 is probably more interesting than the proposition itself. We are proving properties of unrelativized classes by means of relativized classes. Proposition 4.5 makes this possible.

Proposition 4.8. For $n \geq 0$,

a. for almost every pair (T_1, T_2) of tally sets, $BP\cdot\Sigma_n^P =$

$\Sigma_n^P(T_1) \cap \Sigma_n^P(T_2)$;

b. for almost every tally set T, $BP\cdot\Sigma_n^P = REC \cap \Sigma_n^P(T)$,

where REC denotes the class of recursive sets.

The proposition can be interpreted in terms of minimal pairs and minimal sets. We shall not discuss this here. However, by taking $n = 1$ we obtain an interesting equality: $BP\cdot NP = AM = REC \cap NP(T)$ for almost very tally set T. That is, if a set is recursive and belongs to $AM(T)$ for a tally set that is chosen randomly, then with probability 1 this set is in AM.

When **CL** is PH or PSPACE, Proposition 4.5(b) is obvious because of the relativization of Proposition 4.2. But Proposition 4.5(a) becomes interesting: $A \in PH$ if and only if for almost every tally set T, $A \in PH(T)$ if and only if for all tally sets T, $A \in PH(T)$; $A \in PSPACE$ if and only if for almost every tally set T, $A \in PSPACE(T)$ if and only if for all tally sets T, $A \in PSPACE(T)$.

Similar to Proposition 4.8, we have the following:

a. for almost every pair (T_1, T_2) of tally sets, $PH = PH(T_1) \cap PH(T_2)$;
b. for almost every tally set T, $PH = REC \cap PH(T)$;
c. for almost every pair (T_1, T_2) of tally sets, $PSPACE = PSPACE(T_1) \cap PSPACE(T_2)$;
d. for almost every tally set T, $PSPACE = REC \cap PSPACE(T)$.

5. Reducibilities.

Reducibilities play an important role in structural complexity theory. We shall mainly discuss the following types of reducibilities (for a detailed discussion, see [18]).

Set A is *polynomial-time Turing reducible* to set B, written $A \leq_T^P B$, if there is a deterministic polynomial-time oracle machine accepting A relative to oracle B.

Set A is *polynomial-time truth-table reducible* to set B, written $A \leq_{tt}^P B$, if there is a polynomial-time computable function f and a polynomial-time machine M such that for every x,

i) $f(x)$ is a list of words; $f(x) = y_1\#y_2\# \dots \#y_q$, where $\#$ is a marker, and

ii) $x \in A$ if and only if M accepts $\langle x, b_1b_2 \dots b_q\rangle$, where or all i, $1 \leq i \leq q$, b_i = [*if* $y_i \in B$ *then* 1 *else* 0].

Set A is *polynomial-time k-truth-table reducible* to set B (where $k \geq 1$ is a natural number), written $A \leq_{k\text{-}tt}^P B$, if in the above definition, for all x, $f(x)$ is a list of k words.

Set A is *polynomial-time bounded-truth-table reducible* to set B, written $A \leq_{btt}^P B$, if there exists a $k > 0$ such that $A \leq_{k\text{-}tt}^P B$.

Set A is <u>polynomial-time</u> <u>many-one</u> <u>reducible</u> to set B, written $A \leq^P_m B$, if there is a polynomial-time computable function $f: \Sigma^* \to \Sigma^*$ such that for every x, $x \in A$ if and only if $f(x) \in B$.

It is well known that for any two sets A and B, $A \leq^P_m B$ implies $A \leq^P_{1\text{-}tt} B$ but there exist sets C and D such that $C \leq^P_{1\text{-}tt} D$ while $C \not\leq^P_m D$; similarly for each $k > 0$, for any two sets A and B, $A \leq^P_{k\text{-}tt} B$ implies $A \leq^P_{(k+1)\text{-}tt} B$ but there exist sets C and D such that $C \leq^P_{(k+1)\text{-}tt} D$ while $C \not\leq^P_{k\text{-}tt} D$. The same relationship holds for $\leq^P_{btt}$ and $\leq^P_{tt}$, as well as for $\leq^P_{tt}$ and $\leq^P_T$.

For a reducibility $\leq^P_r$, let $P_r(A) = \{B \mid B \leq^P_r A\}$, that is, $P_r(A)$ is the class of sets that is below A under the reducibility $\leq^P_r$. For a class **CL** of languages, let $P_r(\mathbf{CL}) = \cup_{A \in \mathbf{CL}} P_r(A)$. Notice that $P_T(A) = P(A)$ for every set A.

From Proposition 2.2, we know that TALLY is an interesting class of oracles in structure theory. A broader and probably more interesting class is SPARSE, the class of (polynomially) "sparse" sets.

The census function $c_L(n)$ of a language L is defined to be the number of elements of length at most n in L: $c_L(n) = \|\{x \mid x \in L, |x| \leq n\}\|$.

A language is <u>sparse</u> if the census function is bounded by some fixed polynomial.

Obviously, every tally set is sparse but the converse does not hold.

Examples of results and conjectures about sparse sets and complexity classes are the following:

a. [9] Most computer scientists believe that no NP-complete set is sparse.
b. [15] If there is an NP-complete set that is polynomial-time Turing reducible to a sparse set, then $PH = \Sigma_2^P \cap \Pi_2^P$.
c. [20] If there is an NP-complete set that is polynomial-time many-one reducible to a sparse set, then P = NP.
d. [14] NE = E if and only if there is no sparse set in NP - P.
e. [7] The polynomial-time hierarchy collapses if and only if relative to some sparse set the polynomial-time hierarchy collapses if and only if relative to every sparse set the polynomial-time hierarchy collapses.

Motivated by Proposition 2.2, Book and Ko proved:

Proposition 5.1. [10]

a. $P_{tt}(TALLY) = P(TALLY) = P(SPARSE) = P_{tt}(SPARSE) = P/poly$.
b. $P_{btt}(TALLY) \neq P_{tt}(TALLY)$, $P_{tt}(TALLY) \neq P_{btt}(SPARSE)$.

c. $P_m(\text{SPARSE}) \neq P_{1\text{-tt}}(\text{SPARSE}) \neq P_{2\text{-tt}}(\text{SPARSE}) \neq \dots \neq P_{btt}(\text{SPARSE})$.

d. $P_m(\text{TALLY}) = P_{btt}(\text{TALLY})$.

This shows that the behavior of $\leq_m^P, \leq_{1\text{-tt}}^P, \leq_{2\text{-tt}}^P, \dots, \leq_{btt}^P$ with respect to SPARSE and that with respect to TALLY are different. For SPARSE they form an infinite hierarchy, while for TALLY they collapse all the way down. From the viewpoint of TALLY, Proposition 5.1 shows that $\leq_m^P$ is very close to $\leq_{btt}^P$, that $\leq_{tt}^P$ is very close to $\leq_T^P$, and that $\leq_{btt}^P$ is very far from $\leq_{tt}^P$.

Observe that Proposition 5.1(d) asserts that if there is a tally set T_1 such that $A \leq_{btt}^P T_1$ then there is a tally set T_2 such that $A \leq_m^P T_2$.

It is interesting to consider the behavior of a random tally set T in place of SPARSE or TALLY. Does it behave like SPARSE in Proposition 5.1(c) or like TALLY in Proposition 5.1(d)?

Proposition 5.2. [26] For almost every tally set T, $P_m(T) \neq P_{1\text{-tt}}(T) \neq P_{2\text{-tt}}(\text{SPARSE}) \neq \dots \neq P_{btt}(T)$.

Let us pursue another path parallel to Propositions 1.6 and 1.7. The tally version of Proposition 1.6 is given in Propositions 4.5(a) and 4.6(a): $A \in \text{BPP}$ if and only if for almost every tally set T, $A \in P(T)$. The counterpart of Proposition 1.7 is the following:

Proposition 5.3. [26] $A \in P$ if and only if for almost every tally set T, $A \in P_m(T)$.

In this section we defined infinitely many reducibilities that lie between $\leq_m^P$ and $\leq_T^P$. Can we characterize other classes between P and BPP?

Proposition 5.4. [26] $A \in P$ if and only if for almost every tally set T, $A \in P_{btt}(T)$.

It is easy to prove that

Proposition 5.5. For every tally set T, $A \leq_{tt}^P T$ if and only if $A \leq_T^P T$.

Thus, we have

Proposition 5.6. $A \in BPP$ if and only if for almost every tally set T, $A \leq_{tt}^P T$.

This means that the answer to the question posed before Proposition 5.4 is "no."

From Propositions 5.3 and 5.4, we can prove

Proposition 5.7. [26] Let A be any fixed set. For almost every tally set T, $A \leq_{btt}^P T$ implies $A \leq_m^P T$.

Compare Propositions 5.2 and 5.7. Is it the case that Proposition 5.2 asserts that many-one reducibility and bounded truth-table reducibility behave the same with respect to a random tally set, while Proposition 5.7 asserts that they behave differently? Do these results contradict each other? The answer is "no." In Proposition 5.2, we are considering the downwards reduction; the class below a (random) tally set with respect to some reducibility. In Proposition 5.7, we are considering the upwards reduction: the class of (random) tally sets above a fixed set with respect some reducibility. What we can conclude is that downwards reductions show the difference (as in Proposition 5.2) while upwards reductions give us a very different picture.

Similar to Proposition 4.8, we can conclude the following from Proposition 5.2:

Proposition 5.8. Let $r \in \{m, 1\text{-tt}, 2\text{-tt}, \ldots, \text{btt}\}$.

a. For almost every tally set T, $P = REC \cap P_r(T)$.

b. For almost every pair (T_1, T_2) of tally sets, $P = P_r(T_1) \cap P_r(T_2)$.

This means that almost every tally set is so difficult that no recursive set can be polynomial-time many-one reducible to it unless this recursive set is in P. Propositions 5.2, 5.8, and 4.8 show that, in addition to the traditional diagonalization method, randomness is a useful technique for proving the existence of sets with certain properties.

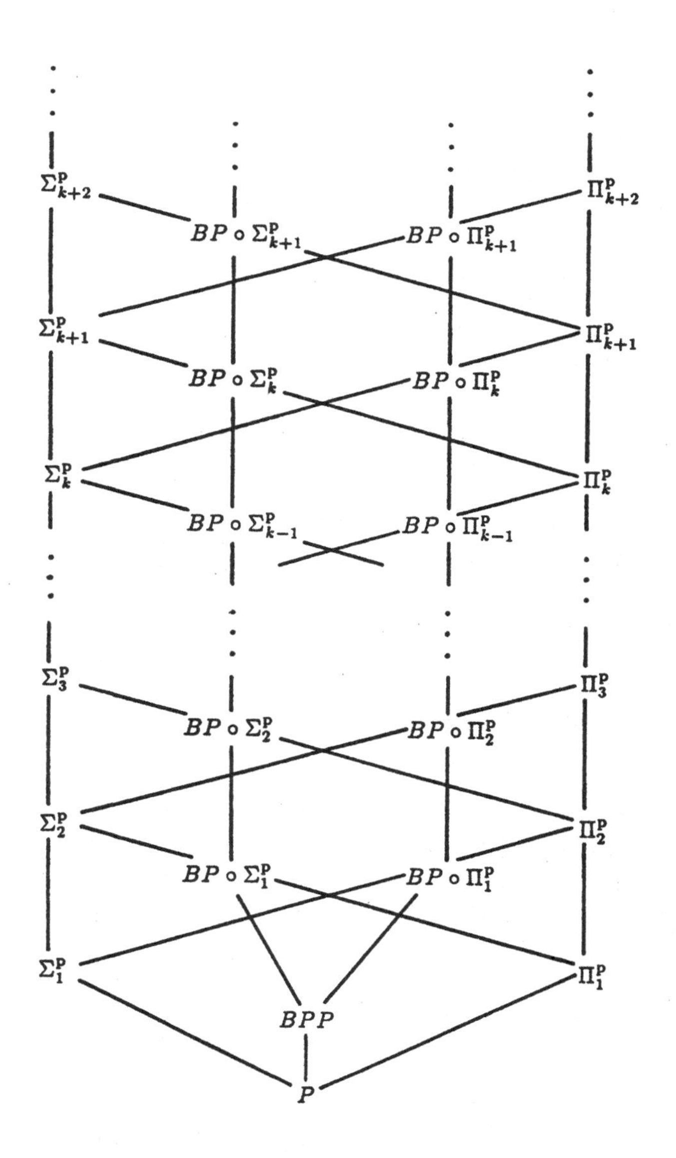

Fig. 1. Polynomial Hierarchy

References

1. Ambos-Spies, K., Randomness, relativizations and polynomial reducibilities, Proc. 1st Conf. Structure in Complexity Theory, Lecture Notes in Computer Science 223 (1986), 23-34.
2. Allender, E. and Rubinstein, R., P-printable sets, SIAM J. Computing 17 (1988), 1193-1202.
3. Allender, E. and Watanabe, O., Kolmogorov complexity and degrees of tally sets, Proc. 3rd IEEE Conf. Structure in Complexity Theory, 1988, 102-111.
4. Babai, L., Trading group theory for randomness, Proc. 17th ACM Symp. Theory of Computing, 1985, 421-429.
5. Babai, L., Random oracle separates PSPACE from the polynomial-time hierarchy, Inf. Proc. Letters 26 (1987/88), 51-53.
6. Baker, T., Gill, J., and Solovay, R., Relativizations of the P = NP question, SIAM J. Computing 4 (1975), 431-442.
7. Balcázar, J., Book, R., and Schöning, U., The polynomial-time hierarchy and sparse oracles, J. Assoc. Ccomput. Mach. 33, 1986, 603-617.
8. Bennett, C. and Gill, J., Relative to a random oracle, P $\neq$ NP $\neq$ co-NP with probability 1, SIAM J. Computing 10 (1981), 96-113.
9. Berman, L. and Hartmanis, J., On polynomial time isomorphisms of complete sets, Proc. 3rd GI Conf. Theoret. Comp. Sci., Lecture Notes in Computer Science 48, 1977, 1-15.
10. Book, R. and Ko, K., On sets reducible to sparse sets, Proc. 2nd IEEE Conf. Structure in Complexity Theory, 1987, 147-154.
11. Cai, J., With probability one, a random oracle separates PSPACE from the polynomial-time hierarchy, Proc. 18th ACM Symp. Theory of Computing, 1986, 21-29.
12. Gill, J., Computational complexity of probabilistic Turing machine, SIAM J. Comput. 6 (1977), 675-695.
13. Hartmanis, J., Solvable problems with conflicting relativizations, Bulletin EATCS 27, 1985.
14. Hartmanis, J., Immerman, N., and Sewelson, V., Sparse sets in NP-P: EXPTIME versus NEXPTIME, Proc. 15th ACM Symp. Theory of Computing, 1983, 382-391.
15. Karp, R. and Lipton, R., Some connections between nonuniform and uniform complexity classes, Proc. 12th ACM Symp. Theory of Computing, 1980, 302-309.

16. Ko, K., Some observations on probabilistic algorithms and NP-hard problems, Inf. Proc. Letters 14 (1982), 39-43.
17. Kurtz, S., On the random oracle hypothesis, Proc. 14th ACM Symp. Theory of Computing, 1982.
18. Ladner, R., Lynch, N., and Selman, A., A comparison of polynomial time reducibilities, Theoret. Comput. Sci. 1 (1972), 103-124.
19. Long, T. and Selman, A., Relativizing complexity classes with sparse oracles, J. Assoc. Comput. Mach. 33, 1986, 618-627.
20. Mahaney, S., Sparse complete sets for NP: solution of a conjecture of Berman and Hartmanis, J. Computer and System Sciences, 25 (1982), 130-143.
21. Orponen, P., Complexity of alternating machines with oracles, Proc. 10th International Colloquium on Automata, Languages and Programming, Lecture Notes in Computer Science 154 (1983), 573-584.
22. Schöning, U., Probabilistic complexity classes and lowness, Proc. 2nd IEEE Conf. Structure in Complexity Theory, 1987, 2-8.
23. Sipser, M., A complexity theoretic approach to randomness, Proc. 15th ACM Symp. Theory of Computing, 1983, 330-335.
24. Stockmeyer, L., The polynomial-time hierarchy, Theoret. Comput. Sci. 3 (1977), 1-22.
25. Tang, S. and Book, R., Separating polynomial-time Turing and truth-table degree of tally sets, Proc. 15th International Colloquium on Automata, Languages and Programming, Lecture Notes in Computer Science 317 (1988), 591-599.
26. Tang, S. and Book, R., Polynomial-time reducibilities and "almost all" oracle sets, Theoret. Comput. Sci., to appear.
27. Tang, S. and Watanabe, O., On tally relativizations of BP-complexity classes, Proc. 3rd IEEE Conf. Structure in Complexity Theory, 1988, 10-18.
28. Wrathall, C., Complete sets and the polynomial-time hierarchy, Theoret. Comput. Sci. 3 (1977), 23-33.
29. Zachos, S. and Furer, M., Probabilistic quantifiers vs. distrustful adversaries, manuscript.

ON ONE-WAY FUNCTIONS†

Osamu Watanabe

Department of Computer Science
Tokyo Institute of Technology
Tokyo 152, Japan

1. Introduction

A function is called *one-way* in general if its inverse is "harder" to compute than the function itself. Recently, one-way functions have received considerable attention because of their practical application as well as their theoretical importance. In the field of cryptography, in particular, one-way functions play an important role; several candidates for one-way functions have been proposed and used there. The subject of this paper is to give an overview of research on computational complexity of one-way functions.

There are many definitions for "one-way function": the definition varies depending on function types, e.g., either one-to-one or many-to-one, and depending on ways to measure "hardness". We mainly consider *one-to-one* and *honest* functions, because these properties are required in most circumstances (the notion of "honest function" is defined later). We use different types of complexity measures based on two different computation models:— the Boolean circuit model and the Turing machine model. In short, we will survey researches on "one-to-

† This paper is the detailed version of a lecture given at the International Symposium on Combinatorial Optimization held at the Nankai Institute of Mathematics in Tianjin, People's Republic of China, August 15 – 18, 1988.

one and honest one-way function" under different criteria of hardness.

It should be mentioned here that, in almost all cases, we have been unable to find a function that is prove to be "one-way". On the other hand, we have several good candidates for one-way functions; for such a candidate f, we conjecture that f satisfies the condition for one-way functions of the corresponding type, and indeed, we can prove that f meets all the requirements in the condition except that f^{-1} has sufficient hardness. By a one-way function, we often mean such a "purported" one-way function.

The most important complexity issue concerning one-way functions is hardness of computing their inverses. There are not a few research programs studying this subject. We will show those on "polynomial time one-way functions"; functions that are almost polynomial time computable, but their inverses are not.

We will characterize computation necessary to compute the inverse of a polynomial time one-way function. Consider the following example:

for every prime numbers p and q, $1 \leq p \leq q$,

$pt(\langle p, q \rangle) = p \cdot q$.

(*NOTE:* pt is undefined for the other inputs.)

It is provable that pt is one-to-one, honest, and almost polynomial time computable; we conjecture that pt^{-1} is not polynomial time computable. Thus, pt is a purported polynomial time one-way function. Now let us discuss the type of computation necessary to evaluate pt^{-1}. Intuitively, pt^{-1} is "nondeterministically" polynomial time computable: for a given n, one can compute $\langle p, q \rangle = pt^{-1}(n)$ in polynomial time by guessing p and q and checking if $p \cdot q = n$. Notice that only one guess leads to the correct value of $pt^{-1}(n)$ (if it exists): namely, the usage of nondeterminism for computing pt^{-1} is more restrictive than the general notion of nondeterminism. Similar observation holds for any polynomial time one-way functions: intuitively, polynomial time nondeterministic computation of some restrictive type, "polynomial time unambiguous computation", is necessary and sufficient for computing the inverse of a polynomial time one-way function. We will clarify this intuition using polynomial time sub-nondeterministic complexity classes. From our observation, we can formally prove, for example, that if a polynomial time one-way function exists, then $P \neq NP$.

One may feel hopeless to obtain some result in the study of hardness of inverting a (purported) one-way function, because even its polynomial time non-invertibility proves $P \neq NP$. Of course, the polynomial time non-invertibility is the most important hardness here; nevertheless, it is not *the only* hardness we should investigate. In structural complexity theory (see, e.g., [37]), we have established several qualitative ways to show "hardness"; we will explain investigations on "structural hardness" of inverting a polynomial time one-way function.

The paper is organized as follows. In Section 2, we explain one example of application of one-way functions to cryptography. In Section 3, we discuss one-way functions defined using circuit complexity measures. Section 4 is for polynomial time one-way functions. We present several definitions for polynomial one-way functions in Section 4.1; complexity classes characterizing computation for their inverses are also defined there. In Section 4.2 we discuss structural hardness of inverting a polynomial time one-way function. In order to shorten the list of references, we often refer one of the latest available paper for each topic and omit the others. We sometimes omit the reference for a well-known fact that can be found in a standard text book such as [4].

In this paper, we use standard notions and notation in computational complexity theory (see, e.g., [4]). In the following we briefly review notions and notation used throughout the paper; some other notions and notation will be defined where they are necessary. Let Σ denote a finite alphabet that includes $\{0,1\}$. By a *string* we mean an element of Σ^*; let ϵ denote the null string. We assume some natural encoding of $\mathbf{N}$, the set of nonnegative integers, over Σ. For any string x, we use $|x|$ to denote the length of x. For any set A and any $n \geq 0$, we use $A^{\leq n}$ and $A^{=n}$ to denote $\{x \in A : |x| \leq n\}$ and $\{x \in A : |x| = n\}$ respectively; let $\|A\|$ denote the number of elements in A. A set S is *sparse* if there is a polynomial p such that $\|S^{\leq n}\| \leq p(n)$ for every $n \geq 0$. We assume some fixed one-to-one and total pairing function that is polynomial time computable and invertible; let $\lambda xy.\langle x, y\rangle$ denote it.

By a *function* we mean a function from Σ^* to Σ^*; a function is not necessarily total. For any function f, we use $Dom(f)$ to denote the domain of f and $Range(f)$ to denote the range of f. For any y, let f^{-1} denote the set $\{x : f(x) = y\}$. A function f is *one-to-one* if $\|f^{-1}(y)\| = 1$ for every $y \in Range(f)$; f is *const-to-one* (resp., *poly-to-one*) if there exists a constant c (resp., a polynomial p) such that $\|f^{-1}(y)\| \leq c$ (resp., $\|f^{-1}(y)\| \leq p(|y|)$) for every $y \in Range(f)$. By an *inverse* of f we mean a function g mapping every $y \in Range(f)$ to some element in $f^{-1}(y)$ (note that $Dom(g) = Range(f)$). Notice that every one-to-one function has a unique inverse; when f is one-to-one, we use f^{-1} also to denote the inverse of f. We usually consider some requirement on the length of function values in order to avoid the case that hardness of computing a function is trivial from the length of its values. A function f is *honest* if for every $x \in Dom(f)$, $|f(x)| \leq p(|x|)$ and $|x| \leq p(|f(x)|)$, where p is some polynomial; f is *length-preserving* if for every $x \in Dom(f)$, $|f(x)| = |x|$.

We use standard notation for complexity classes; by a *complexity class* we mean a class of sets recognized by some resource bounded computation: e.g., LOGSPACE, P, NP. For any Turing machine acceptor M, let $L(M)$ denote the set of strings accepted by M; for any oracle Turing machine acceptor M and any set A, let $L(M, A)$ denote the set of strings accepted by M relative to A. Relativized complexity classes are written, for example, as $\mathrm{P}(A)$.

2. Application of One-Way Functions

The study of one-way functions has consequence in several fields of complexity theory. It has been shown that the existence of some type of one-way function has important implication in structure of complexity classes:— [3], [36], etc. However, even greater importance of one-way functions has been observed in rather practical complexity theory. In the field of cryptography, in particular, one-way functions have been (assumed and) used for many purposes. In this section we show one example of such application. Through the discussion on the example, we will see some of the properties required for "desirable" one-way functions.

In [15] Grollmann and Selman established a framework in which we can discuss complexity theoretical problems concerning "public-key cryptography [11]". Then they discussed the relation between secure public-key cryptosystems and one-way functions; an abstract public-key cryptosystem is designed using some type of one-way function. This is one of the important examples of application. On the other hand, we should note that modern cryptography studies wide variety of topics: the reader will find such topics in, e.g., [38]. Here we choose one of the simplest topics — coin-flipping game by telephone [6] — and explain its relation to one-way functions.

We discuss a way, ***coin-flipping protocol***, to achieve a coin-flipping game by telephone. Although this problem itself is simple, it is a basis of many important problems in cryptography. Indeed, Blum [6], who first considered this game in literature, showed that his coin-flipping protocol can be used to solve many problems:— mental poker, certified mail, etc.

Suppose that two friends, say, Charlie and Patty [26], want to flip a coin by telephone. Here is the simplest way.

(1) Charlie flips a coin first and asks Patty which side he has flipped.
(2) Patty, at the other side of the telephone line, guesses the answer, HEADS or TAILS, and tells him her guess.
(3) Charlie tells Patty the result of his coin-flipping at (1). Patty wins if she did a right guess; Charlie wins otherwise.

Obviously, Charlie can cheat† Patty by telling her his desired side whichever side he flipped at (1). That is, this protocol is not "fair", where a protocol is considered as *fair* if each participant has 50-50 chance to win the game. Here we present a simple fair coin-flipping protocol using some type of one-way function.

Idea of our coin-flipping protocol is easy. We assume a one-way function f on $\{0,1\}^*$ that, intuitively, satisfies the following: (i) f is one-to-one, (ii) f is easy to compute, and (iii) given $f(x)$, it is hard to compute the first bit of x. Using this one-way function f, Charlie and Patty can achieve a coin-flipping game as follows.

(1) Charlie first chooses sufficiently long x randomly, computes $y = f(x)$, and tells y to Patty.

† Well, I think you can believe in Charlie Brown [26].

(2) Patty, who cannot determine the first bit of x from y, guesses the first bit of x and tells it to him.

(3) Charlie shows x to Patty. Patty wins if she guessed x's first bit correctly; Charlie wins otherwise.

In this protocol, Charlie cannot cheat Patty: Patty can check if Charlie cheated her by comparing y and $f(x)$, where y is given at (1), and $f(x)$ is computed by herself from x given at (3). Note that the one-to-one-ness of f is essential here: if, say, $f(1011) = f(0100)$, and Charlie knows it, then he can cheat Patty. On the other hand, property (iii) guarantees that there is no better way for Patty to get the first bit of x than just guessing it. Thus, the probability for Patty to win the game is no more than 50%.

Despite of this simple idea, we face several difficult and important problems once we consider this protocol in more detail. The issue of "almost-everywhere" complexity is one of such problems. In complexity theory we often consider "worst-case" complexity (or, "infinitely often" complexity). However, in practical use of lower bound results, we often need "almost-everywhere" type complexity measure. Consider the above protocol for example. Suppose that, for some f, the first bit of f^{-1} is not polynomial time computable, and thus, f satisfies the above property (iii) from worst-case complexity. Nevertheless, there exists the possibility that some polynomial time algorithm computes the first bit of $f^{-1}(x)$ correctly for many x, and Patty, using this algorithm, can obtain the correct answer at (2) with high probability. What we really need is a function f such that the first bit of f^{-1} is hard to get on all but a small fraction of inputs: namely, property (iii) should be interpreted as "almost-everywhere hardness". We need, say, the following condition for f:

(C1.i) f is a one-to-one and total function from $\{0,1\}^*$ to $\{0,1\}^*$,

(C1.ii) f is polynomial time computable, and

(C1.iii) there exists $\varepsilon << \frac{1}{2}$ such that for every polynomial time computable function g,

$$\frac{\|\{x \in \{0,1\}^n : g(f(x)) = \text{the first bit of } x\}\|}{\|\{0,1\}^n\|} < \frac{1}{2} + \varepsilon,$$

for almost all n.

Then it is clear that our protocol is *almost* fair† if f satisfies (C1).

Properties required in (C1) are often assumed in application. One-to-one-ness is essential in most cases. In cryptography, (i) totality or the property of having an easily recognizable domain is usually required, and (ii) almost-everywhere type hardness is more or less necessary; in [15] the reader will find elaborate explanation on their relation to cryptosystem.

We finish this example by surveying the other coin-flipping protocols. Blum [6] proposed a protocol from the assumption that the function *pt* is almost-everywhere hard to invert. The coin-flipping game itself has not been studied recently. Instead, problems that include this game as a special case are usually discussed: e.g., the "mental porker" problem (see, e.g., [10]).

† It is impossible to make $\varepsilon = 0$; hence, our protocol is not *really* fair. It is an interesting question whether we can construct really fair protocol using f satisfying the above.

3. Circuit Complexity and One-Way Functions

In this section we discuss one-way functions defined using "circuit complexity" to measure hardness. Since there are two types of circuit complexity measure — *size* and *depth*, we will discuss two completely different types of one-way functions.

We consider the standard Boolean circuit model (see, e.g., [4, 17]). Each individual circuit have n input gates, m output gates, AND gates (2 fan-in), OR gates (2 fan-in), and NOT gates (1 fan-in). A circuit can use a special constant gate, "undefined gate", so that it can output *undef*. The fan-out of gates (except output gates) is not bounded. We denote a circuit with n input gates by C_n. For any circuit C, let $size(C)$ and $depth(C)$ denote the size of C and the depth of C respectively; we use C also to denote the function computed by C. For the sake of simplicity, we assume, in this section, every function is appropriate for this circuit model. Namely, we consider only functions from $\{0,1\}^+$ to $\{0,1\}^+$. Furthermore, we assume that for every function f, the length of $f(x)$, if $f(x)$ is defined, is the same for all input x of the same length. Note that these assumptions are not essential restriction. We say that a family $\{C_n\}_{n\geq 1}$ of circuits *computes* f if $f_n = C_n$ for every $n \geq 1$, where f_n denotes the restriction of f on $\{0,1\}^n$. For any set $A \in \{0,1\}^*$, we say that a family of circuits *accepts* A if it computes the characteristic function of A.

We first consider one-way functions defined considering the size of circuits. A family $\{C_n\}_{n\geq 1}$ of circuits is called (a family of) *polynomial size circuits* if there exists some polynomial p such that $size(C_n) \leq p(n)$ for all $n \geq 1$. For any $n \geq 1$, define $size(f_n)$ to be the size of the smallest circuit computing f_n. We say that f *has polynomial size circuits* if there exists a family of polynomial size circuits computing f: in other words,

f has polynomial size circuits if there exists a polynomial p such that $size(f_n) \leq p(n)$ for all $n \geq 1$. The same notion for sets is defined similarly.

Boyack [8][†] proved that there exists a function such that $size(f_n^{-1}) > size(f_n)$ for infinitely many n. This function could be called one-way in a very weak sense. He also introduced the following condition:

(C2.i) f is one-to-one and length-preserving, and

(C2.ii) $size(f_n^{-1})/size(f_n)$ is unbounded.

(*NOTE:* One-to-one-ness may not be required in the original definition.)

It is left open whether a (C2)-type one-way function exists.

In [7] yet a stronger type of condition is considered:

(C3.i) f is one-to-one and honest,

(C3.ii) f has polynomial size circuits, and

(C3.iii) f^{-1} does not have polynomial size circuits.

(*NOTE:* One-to-one-ness is not required in the original definition.)

Note that (C3) is a stronger condition than (C2): i.e., each (C3)-type one-way function yields a function satisfying (C2). Following an argument similar to the one in Section 4.1, we can characterize the existence of (C3)-type one-way functions.

We first review several concepts concerning "nonuniform complexity classes". A function $h : \mathbf{N} \to \Sigma^*$ is called *polynomially length-bounded* if there exists some polynomial p such that $|h(n)| \leq p(n)$ for all $n \geq 0$. For any complexity class $\mathcal{C}$, define $\mathcal{C}$/poly to be the class of languages L such that for some $C \in \mathcal{C}$, some polynomially length-bounded function h, and all $n \geq 0$, we have $L^{=n} = \{x \in \Sigma^n : \langle x, h(n) \rangle \in C\}$.

[†] Due to my fault I could not obtain Boyack's dissertation [8]; the following survey on [8] is from [1].

Define SPARSE to be the class of sparse sets. We use P(SPARSE) and UP(SPARSE) to denote $\bigcup_{S \in \mathrm{SPARSE}} \mathrm{P}(S)$ and $\bigcup_{S \in \mathrm{SPARSE}} \mathrm{UP}(S)$ respectively (see Section 4.1 for the definition of the class UP).

The following relations are known.

Theorem 3.1. For every set $A \in \{0,1\}^*$, the following statements are equivalent.
(1) A has polynomial size circuits.
(2) A is in P/poly.
(3) A is in P(SPARSE), i.e., $A \in \mathrm{P}(S)$ for some sparse set S.

This theorem has the following important interpretation: every family of polynomial size circuits can be simulated by some polynomial time computation relative to some sparse oracle, and vice versa.

We only have a partial relation for UP [31].

Theorem 3.2. For every set A, if A is in UP/poly, then A is in UP(SPARSE).
Remark. From the non-robustness of UP(), i.e., the fact that the unambiguous property of some oracle machine depends on an oracle set (see, e.g., [16]), we have been unable to prove the converse.

In Section 4.1, we show that for every honest function f, $Pref(f) \in$ P if and only if f is polynomial time computable. Noting the interpretation of Theorem 3.1, we can easily modify the proof to obtain the following property.

Proposition 3.3. For every honest function f having polynomial size circuits, $Pref(f) \in \mathrm{P(SPARSE)}$.

It is also easy to modify the proof of Theorem 4.4 and 4.5, thereby showing the following characterization.

Theorem 3.4. A (C3)-type one-way function exists if and only if P(SPARSE) $\neq$ UP(SPARSE).

From Theorem 3.1 and 3.2, we have one sufficient condition for having a (C3)-type one-way function.

Corollary 3.5. If P/poly $\neq$ UP/poly, then a (C3)-type one-way function exists.

Let us consider one example: we show that the function *pt* is a candidate for (C3)-type one-way functions (see Introduction for the definition of *pt*). We can assume that *pt* is appropriate for our circuit model: otherwise, we can modify it easily. Obviously, multiplication of two integers can be achieved by polynomial size circuits. The important point is the complexity of the domain of *pt*: circuits must output $undef$ if *pt* is not defined. Note that the set of prime numbers is in co-R and that co-R $\subseteq$ P/poly. It follows that the domain of *pt* has polynomial size circuits; thus, we have polynomial size circuits computing *pt*. Therefore *pt* is (C3)-type one-way, if factorization does not have polynomial size circuits.

Now we move to another type of one-way function. We consider the depth complexity; more specifically, we use the NC-hierarchy to measure "hardness". Only in this context, we can define a one-way function that is proved to be one-way from no assumption. We will see such a ***provable*** one-way function here.

What follows is a brief review of definition concerning NC (see, e.g., [9]). Let $k \geq 0$ be fixed. A family of circuits $\{C_n\}_{n\geq 1}$ is called ***poly-size log^k-depth*** if $size(C_n) = O(p(n))$ and $depth(C_n) = O(\log^k n)$, where p is some polynomial; it is called ***log-space uniform*** if some log-space bounded deterministic machine produces C_n from input 0^n. A family of circuits is called *NC^k* if it is poly-size $\log^k$-depth and log-space uniform. For any function f, we say that f *has NC^k circuits* if there exists a family of NC^k circuits computing f. The notion of "a set has NC^k circuits" is defined similarly. The notation NC^k is also used to denote the class of sets having NC^k circuits. Let NC denote $\bigcup_{k\geq 0} NC^k$.

The following containments are known:

$$NC^0 \subseteq NC^1 \subseteq \cdots \subseteq NC \subseteq P.$$

Furst, Saxe, and Sipser [13] proved that $NC^0 \neq NC^1$. On the other hand, no strict containment other than this has been shown (see [35] for relativized results). In [13] they proved the separation by showing *parity* has no poly-size constant-depth family of circuits; the function *parity* is defined by

$$\text{for every } x \in \{0,1\}^*, \quad (\text{let } |x| = n \text{ and } x = x_1 x_2 \ldots x_n)$$
$$parity(x) = x_1 \oplus x_2 \oplus \ldots \oplus x_n,$$

where $\oplus$ denotes exclusive-or.

Boppana and Lagarias [7] defined a permutation τ that has NC^0 circuits but whose inverse is as hard to compute as *parity*. Namely, they showed the following type of one-way function:

(C4.i) f is a permutation,

(C4.ii) f has NC^0 circuits, and

(C4.iii) f^{-1} does not have NC^0 circuits.

The function τ is defined as follows:

$$\text{for every } x \in \{0,1\}^*, \quad (\text{let } |x| = n \text{ and } x = x_1 x_2 \ldots x_n)$$
$$\tau(x) = y_1 y_2 \ldots y_n,$$

where $y_i = x_i \oplus x_{i+1}$ $(1 \leq i \leq n-1)$ and

$$y_n = x_1 \oplus x_{\lfloor n/2 \rfloor} \oplus x_n.$$

It is provable that τ is a permutation and that τ is NC^0 computable. Furthermore, we can prove that τ^{-1} indeed has much stronger hardness required in (C4.iii): i.e., *every* bit of τ^{-1} is as hard as *parity*. To see this, consider the first bit of $x = \tau^{-1}(z_1 z_2 ... z_{2n})$ for example: we have $x_1 = parity(z_n ... z_{2n})$.

Håstad [18] obtained a stronger result: he constructed a NC^0 computable permutation whose inverse is as hard as every polynomial time computable function (*Cf.* function *parity* is one of polynomial time computable functions). Let us state it more precisely. For any function f, a set L is called f's last-bit-set if the characteristic function of L is the last bit of f, i.e., for every $x \in \{0,1\}^*$, $x \in L$ if and only if the last bit of $f(x)$ is 1.

Theorem 3.6. There exists a NC^0 computable permutation η such that the last-bit-set of η^{-1} is complete in P by log-space many-one reduction.

Note that if LOGSPACE $\subsetneq$ P, then η^{-1} is not even LOGSPACE-computable. Similarly, noting that NC $\subseteq$ P, we can conjecture that η^{-1} does not have NC^k circuits for any fixed k. On the other hand, some bits of η^{-1} are easy to compute; thus, τ^{-1} is harder than η^{-1} in this sense.

It should be mentioned that we can show "almost-everywhere" type hardness for the above one-way functions. It is known (see, e.g., [17]) that NC^0 circuits cannot compute *parity* for "many" inputs. Hence, NC^0 circuits cannot compute τ^{-1} (resp., η^{-1}) for "many" inputs. Using this idea extensively, Nissan and Wigderson [25] proposed an interesting pseudo-random bit generator.

4. Polynomial Time One-Way Functions

In this section we survey study on polynomial time one-way functions. First we state several definitions for "polynomial time one-way function"; we introduce nondeterministic complexity classes characterizing computation for inverting polynomial time one-way functions. Then we discuss some approaches for investigating hardness of inverse functions for polynomial time one-way functions.

4.1. Definitions of Polynomial Time One-Way Functions

A function is called *polynomial time one-way* in general, if it is "almost" polynomial time computable, but its inverse is not. There are several types of polynomial time one-way functions. Here we consider four conditions each of which determines one-way functions of one type.

First consider the simplest definition. A function f is called *strictly one-to-one one-way* if it satisfies the following:

(C5.i) f is one-to-one and honest,

(C5.ii) f is polynomial time computable, and

(C5.iii) f^{-1} is not polynomial time computable.

We can define the notion of "strictly const-to-one one-way" by relaxing (C5.i). This generalization is not essential in some sense: we have the following proposition [33].

Proposition 4.1. A strictly one-to-one one-way function exists if and only if a strictly const-to-one one-way function exists.

On the other hand, the following generalization [2] turned out to provide an interesting complexity class:

(C6.i) f is poly-to-one and honest,

(C6.ii) f is polynomial time computable, and

(C6.iii) no inverse of f is polynomial time computable.

In spite of the simple and quite natural definition, we have been unable to find a good example (i.e., a good candidate) for one-way functions of these types; a rather artificial candidate is reported in [15].

Looking at condition (C5), one may claim that if "factorization" is not polynomial time computable as believed, then the function *pt* is strictly one-to-one one-way: namely, *pt* is a good candidate for strictly one-to-one one-way functions. Unfortunately, we have not been able to claim so, since we do not know if *pt* is polynomial time computable. Let us review the definition of "polynomial time computable function".

A function f is called ***polynomial time computable*** if there exists a polynomial time deterministic Turing transducer M such that

(i) $(\forall x \in Dom(f))[$ M on x outputs $f(x)$ $]$; and

(ii) $(\forall x \notin Dom(f))[$ M on x does not halt $]$.

Intuitively, (i) states that the computation of "getting values of f" is polynomial time, and (ii) states that the computation of "recognizing the domain of f" is polynomial time. Although the above definition is standard, a function satisfying (i) is sometimes regarded as "polynomial time computable"; we will call such a function a ***pseudo polynomial time computable function***. The following characterizations are important, though their proof is immediate from the definition.

Proposition 4.2.

(1) A function is pseudo polynomial time computable if and only if it has a polynomial time computable extension.

(2) For every pseudo polynomial time computable function f, it is polynomial time computable if and only if $Dom(f)$ is in P.

Define *times* by $times(\langle n, m\rangle) = n \cdot m$, for every pair of integers n

and m. It is clear that *times* is a polynomial time computable extension of *pt*: i.e., *pt* is pseudo polynomial time computable. However, we have been unable to prove that it is indeed polynomial time computable, because we do not know so far if $Dom(pt)$ is in P. On the other hand, although *pt* has a polynomial time computable extension, we have been unable to find a *one-to-one* and polynomial time computable extension. All purported one-way functions proposed in cryptography have a similar problem (see [15] for the other examples).

Here we introduce two types of one-way functions, a "randomized one-way function" and an "extensible one-way function", that are more appropriate to discuss well-known purported one-way functions.

A function f is called ***randomized polynomial time computable*** if there exists a polynomial time randomized Turing transducer M such that for some $\varepsilon < \frac{1}{2}$, we have

(i) $(\forall x \in Dom(f))[\ \Pr\{M \text{ on } x \text{ does not output } f(x)\} < \varepsilon\ ]$, and

(ii) $(\forall x \notin Dom(f))[\ \Pr\{M \text{ on } x \text{ halts}\} < \varepsilon\ ]$.

A function f is called ***randomized one-way*** if it satisfies the following condition:

(C7.i) f is one-to-one and honest,

(C7.ii) f is randomized polynomial time computable, and

(C7.iii) f^{-1} is not randomized polynomial time computable.

For example, *pt* is one-to-one and honest; it is randomized polynomial time computable, since its domain is in BPP. Hence, *pt* is randomized one-way if its inverse is not randomized polynomial time computable.

A function f is called ***extensible one-way*** if it satisfies the following:

(C8.i) f is one-to-one and honest,

(C8.ii) f has a polynomial time computable extension g such that for every $y \in Range(f)$, $g^{-1}(y) = \{f^{-1}(y)\}$, and

(C8.iii) f^{-1} has no polynomial time computable extension.

It is easy to see that well-known purported one-way functions considered in cryptography are (or yield) extensible one-way functions if their inverses are indeed as hard as expected. For example, *pt* satisfies (C8-i) and (C8-ii): *times* witnesses of *pt* satisfying (C8-ii). Thus, *pt* is extensible one-way if pt^{-1} is not pseudo polynomial time computable, or no polynomial time machine computes pt^{-1} correctly on $Range(pt^{-1})$. Grollman and Selman pointed out close relation between extensible one-way functions and public-key cryptosystems: they proved [15; Theorem 11] that if there exists an extensible one-way function whose domain is in NP, then one can construct an abstract public-key cryptosystem that cannot be cracked in polynomial time.

Many important concepts in computational complexity theory have been established considering "decision problems". In order to use such concepts, we need to discuss our problem in the context of decision problems†, although we are interested in the problems of "evaluating functions". Here we investigate the relation between decision problems and evaluation problems. We show close relation between them and define classes of sets having hardness similar to the one for inverting a polynomial time one-way function; thereby, we can discuss hardness of inverse functions in the context of decision problems.

The notion of "prefix set of function values" is important to relate evaluation problems to decision problems. For any function f, define a

† We should notice that this approach has limitation, since "decision" and "evaluation" are not exactly the same. Recently, some researchers have obtained interesting results investigating more directly problems of evaluating functions (see, e.g., [19, 5]).

prefix set of f, $Pref(f)$, as follows:

$$Pref(f) = \{\langle x, w\rangle : x \in Dom(f) \wedge w \text{ is a prefix of } f(x)\}.$$

We can observe that both f and $Pref(f)$ have similar hardness: the following proposition states one consequence of this similarity.

Proposition 4.3. For every honest function f, $Pref(f)$ is in P if and only if f is polynomial time computable.

Proof. (*If Part*) The proof follows immediately noting that one can check whether a given input $\langle x, w\rangle$ is in $Pref(f)$ if he knows $f(x)$.
(*Only-If Part*) The following proof is from [29]. For a given x, one can determine $f(x)$ bit by bit from the left to the right asking queries to $Pref(f)$. Note that $|f(x)| \le p(|x|)$ by some polynomial p since f is honest. Thus, the assumption $Pref(f) \in \mathrm{P}$ implies that f is polynomial time computable. □

Remark. The above proof reveals much closer relation than the one stated in the proposition. Intuitively, the proof states that $Pref(f)$ is polynomial time computable using f as an oracle function and that f is polynomial time computable using $Pref(f)$ as an oracle set; thus, both f and $Pref(f)$ are regarded as members of the same $\le_{\mathrm{T}}^{\mathrm{P}}$-degree. Hence, properties preserved under $\le_{\mathrm{T}}^{\mathrm{P}}$-reducibility are shared by f and $Pref(f)$ (see the second topics in Section 4.2). On the other hand, f and $Pref(f)$ may have different types of hardness when we investigate their hardness in more detail (see the third topic in Section 4.2).

In Introduction we intuitively explained that the inverse of a polynomial time one-way function is nondeterministically computable within polynomial time. Now we clarify this intuition. We define sub-nondeterministic complexity classes UP, FewP, UP(BPP), and $\mathcal{UP}$ that respectively characterizes computation for inverting the corresponding type of one-way function.

We first define classes UP [29], FewP [2], and UP(BPP). We say that a nondeterministic machine M is ***unambiguous on*** x if there exists ***at most*** one accepting path in the execution of M on input x; M is an ***unambiguous machine*** if M is unambiguous on every input. We extend this notion as follows: a ***few-accepting-path machine*** is a nondeterministic machine having ***at most*** $p(|x|)$ many accepting paths for every input x, where p is some polynomial. The concept of "unambiguous machine" is relativized as follows: for any set A, a nondeterministic machine M is ***unambiguous relative to*** A if M has at most one accepting path relative to A for every input. Now define three complexity classes as follows:

$$\begin{aligned}
\mathrm{UP} &= \{L(M) : M \text{ is polynomial time and unambiguous}\},\\
\mathrm{FewP} &= \{L(M) : M \text{ is polynomial time and few-accepting-path}\},\\
\mathrm{UP(BPP)} &= \{L(M,A) : A \in \mathrm{BPP} \text{ and } M \text{ is polynomial time}\\
&\qquad\qquad \text{and unambiguous relative to } A\}.
\end{aligned}$$

The following relation is clear from the definition:

$$\mathrm{P} \subseteq \mathrm{UP} \subseteq \mathrm{FewP} \subseteq \mathrm{NP} \quad \text{and} \quad \mathrm{UP} \subseteq \mathrm{UP(BPP)}.$$

Next we define a class $\mathcal{UP}$ of unambiguous promise problems. Even and Yacobi [12] introduced "promise problem": A pair of sets (Q,R) is called a ***promise problem***, where Q and R are regarded as a ***promise*** and a ***property*** respectively. A set X is called a ***solution*** of a promise problem (Q,R) if $(\forall x \in Q)[\ x \in R \leftrightarrow x \in X\ ]$. We say that a promise problem (Q,R) is ***polynomial time solvable*** if there is a solution of (Q,R) that is in P. Here we use one type of promise problem, "unambiguous promise problem", for investigating extensible one-way functions. A promise problem (Q, R) is called *unambiguous* if there exists a polynomial time nondeterministic machine M accepting R such that M on Q is unambiguous. We use $\mathcal{P}$ and $\mathcal{UP}$ to denote the class of polynomial time solvable promise problems and the class of unambiguous promise problems respectively. Note that every unambiguous promise problem has a solution in NP, and hence, $\mathcal{P} \subseteq \mathcal{UP}$.

Now we show the relation between classes UP, FewP, UP(BPP), and $\mathcal{UP}$ and the corresponding type of one-way functions. First we observe that, for example, computing the inverse of a (C5)-type function is no harder than recognizing a set in UP; the following proposition is its consequence.

Proposition 4.4.

(1) If a (C5)-type one-way function exists, then P $\neq$ UP.

(2) If a (C6)-type one-way function exists, then P $\neq$ FewP.

(3) If a (C7)-type one-way function exists, then BPP $\neq$ UP(BPP).

(4) If a (C8)-type one-way function exists, then $\mathcal{P} \neq \mathcal{UP}$.

Sketch of Proof. We prove (1) and (4). The key point for proving (1) is the fact that for every f satisfying (C5-i) and (C5-ii), $Pref(f^{-1})$ is in UP. Thus, if f is (C5)-type one-way; then $Pref(f^{-1})$ must belong to UP $-$ P (see also Proposition 4.3). We can prove (2) and (3) similarly.

Now suppose that f is a function satisfying (C8). Let g be its polynomial time computable extension witnessing (C8-ii); we can assume that g is honest. Then it is easy to see that $(Range(f), Pref(g^{-1}))$ is an unambiguous promise problem. Let X be any solution of this promise problem: i.e., X gives a correct answer for every y in $Range(f)$. Recall the (*Only-If Part*) proof of Proposition 4.3; following a similar argument, we can prove that some extension of f^{-1} is polynomial time computable using X as an oracle. Hence, $(Range(f), Pref(g^{-1}))$ has no solution in P. □

Remark. We have proved much stronger relation by the above proof. For example, we showed that for every function f satisfying (C5.i) and (C5.ii), $Pref(f^{-1})$ is in UP: namely, UP computation is sufficient to obtain inverse values of a (C5)-type one-way function.

Next we investigate the converse relation;

Proposition 4.5.

(1) If P $\neq$ UP, then a (C5)-type one-way function exists.

(2) If P $\neq$ FewP, then a (C6)-type one-way function exists.

(3) If BPP $\neq$ UP(BPP), then a (C7)-type one-way function exists.

(4) If $\mathcal{P} \neq \mathcal{UP}$, then a (C8)-type one-way function exists.

Sketch of Proof. We present the proof of (1) stated in [15, 20]; the others are proved similarly. Let L be a set in UP $-$ P. Then there exists a polynomial time unambiguous machine M accepting L. We can assume some polynomial p such that every computation of M on an input of length n is encoded by a string in $\{0,1\}^{=p(n)}$. Our desired function f_L is defined as follows:

$$f_L(\langle x, w\rangle) = \begin{cases} \langle x, 0^{p(|x|)}\rangle, & \text{if } w \text{ is an accepting path of } M \text{ on } x; \\ \langle x, w\rangle, & \text{otherwise.} \end{cases}$$

Then f_L satisfies (C5-i) and (C5-ii); the one-to-one-ness of f_L is due to the unambiguous property of M. Note that for every x, $x \in L \leftrightarrow f_L^{-1}(\langle x, 0^{p(|x|)}\rangle)$ is defined $\leftrightarrow \langle\langle x, 0^{p(|x|)}\rangle, \epsilon\rangle \in Pref(f_L^{-1})$, where ϵ denotes the null string. Thus, $Pref(f_L^{-1}) \notin$ P: i.e., f_L^{-1} is not polynomial time computable. □

Remark. We have proved in the above that for every L in UP, there exists some f_L satisfying (C5-i) and (C5-ii) such that $L \leq_{\mathrm{m}}^{\mathrm{P}} Pref(f_L^{-1})$: namely, every set in UP is no harder than computing the inverse of a (C5)-type one-way function.

We close this section pointing out the relation between polynomial time one-way functions and the class NP. It follows from the definition that each one of the separations, P $\neq$ UP, P $\neq$ FewP, or $\mathcal{P} \neq \mathcal{UP}$, implies P $\neq$ NP. From the study of the class BPP, we also have that if BPP $\neq$ UP(BPP), then P $\neq$ NP. Hence, the existence of a polynomial time one-way function of any type considered above immediately proves P $\neq$ NP.

4.2. Hardness of Computing Inverse Functions

Here we survey investigations on hardness of computing the inverse of a polynomial time one-way function.

In Section 4.1 we saw that the affirmative answer to our fundamental question — 'Is there any polynomial time computable function whose inverse is not polynomial time computable?' — solves long standing open problems such as P =? NP. Knowing this, one may claim, 'Prove P $\neq$ NP first; otherwise, discussion on hardness of inverse functions would be nonsense!' Another would say, 'Research on hardness of inverse functions is too difficult for *me*, because even the problem of showing polynomial time non-invertibility is hopeless to me!' Of course, the polynomial time non-invertibility is the most important "hardness" for polynomial time one-way functions. However, it is not *the only* hardness we should investigate; showing polynomial time non-invertibility is not *the only* result meaningful. We can think of research programs on hardness of inverse functions which are "feasible" but still important. Here we survey three research programs and results there.

<u>Lower Bound</u>

The first research area is due to Allender. In his dissertation [1] he asked the following question.

Question 1. How hard must a function be so that its inverse is not polynomial time computable?

In other words, he asked lower bounds of computing polynomial time one-way functions.

The following is one of Allender's observation concerning this question (see [1] for related results).

Theorem 4.6. Let f be an honest (not necessarily one-to-one) function computed by a pushdown machine with a one-way input tape and logspace work tape. Then f has an polynomial time computable inverse.

In the next two topics, we discuss "hardness" by methods established in structural complexity theory. We first review some of the basic notions.

The notion of "reducibility" provides the way of comparing difficulty between two problems. Here we use reducibilities so called "polynomial time reducibility". Note that there are many variations of "polynomial time reducibility" (see, e.g., [21]). We assume that the reader knows the typical ones:— $\leq_{\mathrm{m}}^{\mathrm{P}}$-reducibility and $\leq_{\mathrm{T}}^{\mathrm{P}}$-reducibility. Besides them we need the following reducibilities.

Definition 4.1.

(1) A *1-tt-function* is a total function whose value is always a pair of string and a Boolean predicate in $\{id, \neg\}$, where id and $\neg$ respectively indicates the identity Boolean function and the negation function. For any A and B, A is ***polynomial time 1-tt-reducible*** to B ($A \leq_{\text{1-tt}}^{\mathrm{P}} B$) if there is a polynomial time computable 1-tt-function f such that for every x (let $f(x) = \langle y, \alpha \rangle$), $x \in A \leftrightarrow \alpha(y \in B) = \mathbf{true}$.

(2) For any A and B, A is *polynomial time strongly nondeterministic Turing reducible* to B ($A \leq_{\mathrm{T}}^{\mathrm{SN}} B$) if $A \in \mathrm{NP}(B) \cap \text{co-NP}(B)$.

(3) For any A and B, A is *polynomial time randomized reducible* to B ($A \leq_{\mathrm{T}}^{\mathrm{BPP}} B$) if $A \in \mathrm{BPP}(B)$, i.e., there exists a polynomial time probabilistic oracle machine M such that (i) M has bounded error probability relative to B, and (ii) $A = L(M, B)$.

A set A is called *NP-hard* if every set in NP is polynomial time reducible to it; A is called *NP-complete* if it is NP-hard and belongs to NP. It should be mentioned that each type of polynomial time reducibility is conjectured to yield different type of "NP-hardness" and "NP-completeness" [32]. In usual context, e.g., [14], "NP-complete" means NP-complete under $\leq_{\mathrm{m}}^{\mathrm{P}}$-reducibility.

NP-Completeness

Intuitively, the inverse of any polynomial time computable function is NP computable. Thus, for a given polynomial time (purported) one-way function f, it is natural to ask the following question.

Question 2. How hard is it in NP to compute f^{-1}?

A standard approach for this type of question is to show "NP-completeness" (or more generally "NP-hardness"): an NP-complete problem can be regarded as the most difficult problem in NP, because by NP-completeness one can guarantee that the problem is not polynomial time solvable unless *every* NP problem is polynomial time solvable. We follow this approach and construct a purported extensible one-way function *one-sat* whose inverse is NP-hard to compute, where $\leq_{\mathrm{T}}^{\mathrm{BPP}}$-reducibility is used for polynomial time reducibility.

We assume some natural encoding of Boolean formulas over Σ; in the following a "Boolean formula" means a string encoding the formula.

Define a set SAT to be the set of all satisfying Boolean formulas, and define 1SAT to be the set of all Boolean formulas that have *at most* one satisfying assignment. It is clear that a promise problem (1SAT, SAT) is unambiguous (see Section 4.1 for the definition): a natural nondeterministic machine for SAT witnesses it. Valiant and Vazirani [30] proved the following.

Theorem 4.7. Every solution of promise problem (1SAT, SAT) is NP-hard under $\leq_{\mathrm{T}}^{\mathrm{BPP}}$-reducibility.

We can assume some polynomial p such that for every Boolean formula x, its assignment is encoded by the string of length $p(|x|)$. Now define *one-sat* by

for every x and $y \in \Sigma^*$,

$$\mathit{one\text{-}sat}(\langle x, y\rangle) = \begin{cases} \langle x, 0^{p(|x|)}\rangle, & \text{if } x \text{ is in 1SAT} \cap \text{SAT, and} \\ & y \text{ is a satisfying assignment of } x; \\ \langle x, y\rangle, & \text{otherwise.} \end{cases}$$

Then it is clear that *one-sat* satisfies (C8-i) and (C8-ii).

Requirement (C8-iii) concerns hardness of the easiest extension of $\mathit{one\text{-}sat}^{-1}$. From the above theorem, we can claim that every extension of $\mathit{one\text{-}sat}^{-1}$ is NP-hard to compute; formally, we have the following corollaries.

Corollary 4.8. For every extension g of $\mathit{one\text{-}sat}^{-1}$, $\mathit{Pref}(g)$ is NP-hard under $\leq_{\mathrm{T}}^{\mathrm{BPP}}$-reduciblity.

Remark. Since the "NP-hardness" notion is defined for sets, we state our claim by the NP-hardness of $\mathit{Pref}(g)$ instead of g itself. Recall that we can consider both $\mathit{Pref}(g)$ and g have similar hardness (see Proposition 4.2).

Corollary 4.9. If BPP $\neq$ NP, then *one-sat* is extensible one-way, i.e., *one-sat* satisfies (C8).

Remark. In [30] they proved Theorem 4.7 considering more restrictive reducibility, and thus, we have the following stronger result: if R $\neq$ NP, then *one-sat* is extensible one-way.

The above observation shows that $one\text{-}sat^{-1}$ has desirable hardness; nevertheless, *one-sat* is not considered as a "good" one-way function, since its domain is also difficult (see the discussion in Section 2). One way to obtain such a good one-way function is to restrict *one-sat* so that the domain has reasonable complexity. Notice that while restricting *one-sat*, we must not lose the hardness of $one\text{-}sat^{-1}$; the following story is instructive in this sense. Merkle and Hellman [24] proposed a purported one-way function based on Knapsack Problem, one of NP-complete problems. They considered some restriction on Knapsack Problem so that the function can possess the "trap-door" property. However, because of this restriction, the function was proved to be polynomial time invertible later by Shamir [28].

In our approach to Question 2, we want to know whether f^{-1} is NP-hard to compute; but how? The observation in [33] provides the following way (see [33] for the related results).

Theorem 4.10. For every purported strictly one-to-one one-way function f (i.e., a function f satisfying (C5-i) and (C5-ii)), $Pref(f^{-1})$ is NP-hard if and only if $Range(f)$ is NP-hard, where NP-hardness is considered under $\leq_{\mathrm{T}}^{\mathrm{SN}}$-reducibility.

Remark. Again, we consider $Pref(f^{-1})$ instead of f^{-1}. For the sake of simplicity, we state the theorem only for (C5) condition; but, a similar theorem also holds considering (C7) and (C8).

Therefore, for a given function f, we can decide if f^{-1} is NP-hard to compute by investigating the NP-hardness of its range. Note that our proof technique requires one-to-one-ness; a similar characterization for many-to-one functions is left open.

Polynomial Time Reducibility to a Sparse Set

Let us discuss in the context of set recognition problems: we investigate "intractability" of the classes UP, FewP, UP(BPP), and $\mathcal{UP}$. Furthermore, let us choose one of such classes, say UP, for the sake of simplicity. Although we have been unable to prove P $\neq$ UP, we can ask the following question.

Question 3. Suppose that P $\subsetneq$ UP. Is it the case that there exists some set in UP that has stronger sense of intractability than polynomial time non-computability?

We investigate this question, while interpreting "stronger intractability" as "polynomial time reducibility to no sparse set".

In general, every set that is polynomial time reducibility to a sparse set has some sort of feasible algorithm, even though it may not have a polynomial time algorithm (see, e.g., [23, 32]). In other words, the notion of "polynomial time reducibility to no sparse set" yields a stronger notion of intractability than polynomial time non-computability. This notion has been studied by several researchers (see, e.g., [23]). For example, Mahaney [22] proved that if P $\neq$ NP, then NP has a set $\leq^{\rm P}_{\rm m}$-reducible to no sparse set: namely, if NP has a set not in P, it indeed has a set with stronger sense of intractability. We ask if a similar theorem holds for the class UP.

While $\leq^{\rm P}_{\rm m}$-reducibility is considered in Mahaney's Theorem, we investigate the question considering more general reducibility, $\leq^{\rm P}_{\rm 1\text{-}tt}$-

reducibility. This generalization makes it possible to discuss the non-existence of some approximation algorithms, i.e., "p-close approximations". A set A is said to ***have a p-close approximation*** (see, e.g., [27]) if there is a polynomial time algorithm M such that $A \triangle L(M)$ is sparse, where $\triangle$ denotes the symmetric difference operation. It is easy to show that every set $\leq^{\mathrm{P}}_{1\text{-tt}}$-reducible to no sparse set is not $\leq^{\mathrm{P}}_{1\text{-tt}}$-reducible to a set with p-close approximation [27]. Thus, if a set is not $\leq^{\mathrm{P}}_{1\text{-tt}}$-reducible to a sparse set, then not only is it intractable by not having a polynomial time algorithm, but also by not ($\leq^{\mathrm{P}}_{1\text{-tt}}$-reducible to any set) having a p-close approximation.

Let us state our answer to Question 3. Watanabe [34] proved the following theorem.

Theorem 4.11. If $\mathrm{P} \neq \mathrm{UP}$, then there exists a set in UP that is $\leq^{\mathrm{P}}_{1\text{-tt}}$-reducible to no sparse set.

Corollary 4.12. If $\mathrm{P} \neq \mathrm{UP}$, then there exists a set in UP that is not $\leq^{\mathrm{P}}_{1\text{-tt}}$-reducible to any set having a p-close approximation.

Remark. Similar results hold for UP(BPP) and $\mathcal{UP}$; on the other hand, we have slightly weaker results for FewP.

Now discuss the consequence of Theorem 4.11 to the study of hardness of inverse functions. We first define the concepts of "approximation" and "reducibility" appropriately. For any function f, we say that ***f has a p-close approximation*** if there exists a polynomial time computable function g such that $\{x \in Dom(f) : f(x) \neq g(x)\}$ is sparse. For any functions f and g, we say that f is $\leq^{\mathrm{PF}}_{\mathrm{inv}}$-reducible to g if there exist total functions h_1 and h_2 such that (i) both h_1 and h_2 are polynomial time computable, (ii) h_2 has a polynomial time computable inverse, and (iii) $f = \lambda x.h_1(x, g(h_2(x)))$.

From Theorem 4.11, we have the following observation[†].

Corollary 4.13. If a strictly one-to-one one-way function exists, then there exists a strictly one-to-one one-way function f such that f^{-1} is not $\leq_{\mathrm{inv}}^{\mathrm{PF}}$-reducible to any function having a p-close approximation.

Remark. In spite of the intimate relation between UP and strictly one-to-one one-way functions, the intractability of UP shown in Theorem 4.11 has no corresponding natural interpretation for hardness of inverting a strictly one-to-one one-way function; neither this corollary fully reflects Theorem 4.11.

Notice that Corollary 4.13 exhibits "almost-everywhere" type hardness, although the hardness is of very weak type.

Acknowledgment

I would like to thank to Mr. Y. Shinoda at Tokyo Institute of Technology for his help when preparing this manuscript by TEX.

[†] There is an error in [32; Theorem 5.7]; it should be replaced by the one stated here.

References

[1] Allender, E., Invertible Functions, Ph.D. Dissertation, Georgia Institute of Technology (1985).

[2] Allender, E., The complexity of sparse sets in P, in Proc. 1st Structure in Complexity Theory Conference, Lecture Notes in Computer Science 223, Berlin, Springer-Verlag (1986), 1-11.

[3] Allender, E., Some consequences of the existence of pseudorandom generators, SIAM J. Comput. (1988), to appear.

[4] Balcázar, J., Díaz, J., and Gabarró, J., Structural Complexity I, EATCS Monographs on Theoretical Computer Science, Berlin, Springer-Verlang (1988).

[5] Beigel, R., A structural theorem that depends quantitatively on the complexity of SAT, in Proc. 2nd Structure in Complexity Theory Conference, IEEE (1987), 28-32.

[6] Blum, M., Coin flipping by telephone, in Proc. IEEE Spring COMPCON, IEEE (1982), 133-137.

[7] Boppana, R. and Lagarias, J., One-way functions and circuit complexity, Inform. and Comput., 74 (1987), 225-240.

[8] Boyack, S., The Robustness of Combinatorial Measures of Boolean Matrix Complexity, Ph.D. Dissertation, Massachusetts Institute of Technology (1985).

[9] Cook, S., A taxonomy of problems with fast parallel algorithms, Inform. and Control, 64 (1985), 2-22.

[10] Crépeau, C., A zero-knowledge poker protocol that achieves confidentiality of the players' strategy, or how to achieve an electronic poker face, in Proc. Advances in Cryptology — CRYPTO 86, Lecture Notes in Computer Science 263, Berlin, Springer-Verlag (1987), 239-247.

[11] Diffie, W. and Hellman, M., New directions in cryptography, IEEE Trans. Inform. Theory, IT-22 (1976), 644-654.

[12] Even, S. and Yacobi, Y., Cryptocomplexity and NP-completeness, in "Proc. 8th International Colloquium on Automata, Languages and Programming", Lecture Notes in Computer Science 223, Berlin, Springer-Verlag (1980), 195-207.

[13] Furst, M., Saxe, J., and Sipser, M., Parity, circuits, and the polynomial time hierarchy, Math. Syst. Theory, 17 (1984), 13-27.

[14] Garey, M. and Johnson, D., Computers and Intractability: A Guide to the Theory of NP-Completeness, San Francisco, Freeman (1979).

[15] Grollmann, J. and Selman, A., Complexity measures for public-key cryptosystems, SIAM J. Comput., 17 (1988) 309-335.

[16] Hartmanis, J. and Hemachandra, L., One-way functions, robustness, and the non-isomorphism of NP-complete sets, in Proc. 2nd Structure in Complexity Theory Conference, IEEE (1987), 160-174.

[17] Håstad, J., Computational Limitations for Small-Depth Circuits, Cambridge, The MIT Press (1987).

[18] Håstad, J., One-way permutations in NC^0, Inform. Process. Lett., 28 (1987/88), 153-155.

[19] Krentel, M., The complexity of optimization problems, in Proc. 18th ACM Ann. Sympos. on Theory of Computing, ACM (1986), 69-76.

[20] Ko, K., On some natural complete operators, Theoret. Comput. Sci., 37 (1985), 1-30.

[21] Ladner, R., Lynch N., and Selman, A., A comparison of polynomial time reducibilities, Theoret. Comput. Sci., 1 (1975), 103-123.

[22] Mahaney, S., Sparse complete sets for NP: solution of a conjecture of Berman and Hartmanis, J. Comput. Syst. Sci. 25 (1982), 130-143.

[23] Mahaney, S., Sparse sets and reducibilities, in Studies in Complexity Theory, London, Pitman, (1986) 63-118.

[24] Merkle, R. and Hellman, M., Hiding information and signatures in trapdoor knapsacks, IEEE Trans. Information Theory, IT-24 (1978), 525-530.

[25] Nissan, N. and Wigderson, A., Hardness vs. randomness, in Proc. 30th Ann. Sympos. Foundation of Computer Science, IEEE (1988), 2-11.

[26] Schulz, M. and Millward, J., The Peanuts Trivia & Reference Book, New York, Henry Holt and Company, Inc. (1986).

[27] Schöning, U., Complete sets and closeness to complexity classes, Math. Syst. Theory, 19 (1986), 29-41.

[28] Shamir, A., A polynomial time algorithm for breaking the basic Merkle-Hellman cryptosystem, in Proc. 23th Ann. Sympos. on Foundations of Computer Science, IEEE (1982), 145-152.

[29] Valiant, L., Relative complexity of checking and evaluating, Inform. Process. Lett., 5 (1976), 20-23.

[30] Valiant, L. and Vazirani, V., NP is as easy as detecting unique solutions, Theoret. Comput. Sci., 47 (1986), 85-93.

[31] Watanabe, O., On intractability of finite sets, unpublished manuscript.

[32] Watanabe, O., Polynomial time reducibility to a set of small density, in Proc. 2nd Structure in Complexity Theory Conference, IEEE (1987), 138-146.

[33] Watanabe, O., On the difficulty of one-way functions, Inform.

Process. Lett., 27 (1988), 151-157.

[34] Watanabe, O., On $\leq^{P}_{1\text{-tt}}$-sparseness and nondeterministic complexity classes, in Proc. 15th International Colloquium on Automata, Languages and Programming, Lecture Notes in Computer Science 317, Berlin, Springer-Verlag (1988), 697-709.

[35] Wilson, C., Relativized NC, Math. Syst. Theory, 20 (1987), 13-29.

[36] Young, P., Juris Hartmanis: fundamental contributions to isomorphism problems, in Proc. 3rd Structure in Complexity Theory Conference, IEEE (1988), 138-154.

[37] Proc. 3rd Structure in Complexity Theory Conference, IEEE (1988).

[38] Proc. Advances in Cryptology — CRYPTO '87, Lecture Notes in Computer Science 293, Berlin, Springer-Verlag (1988).

A NEW LOWER BOUND FOR PARITY CIRCUITS

Du Dingzhu

Institute of Applied Mathematics
Academia Sinica, Beijing, China

Abstract: We show a new lower bound for parity circuits, which is better than previous ones by Furst, Saxe and Sipser, Yao, Hastad, and Moran.

1. Introduction

A Boolean circuit is a digraph without (directed) cycle and each vertex is either a logical gate or a Boolean variable. An arc from the vertex A to the vertex B represents that the output of A is sent to B as one of its inputs. For each logical gate, its indegree is also called its fan-in. All Boolean variables are inputs of the circuit. The output of the circuit is chosen at a specific gate of outdegree 0. We assume that only one such gate exists in the circuit.

In this paper, we consider only the levelable circuits consisting of gates $\vee$, $\wedge$ (and implicitly $\neg$) as shown in Figure 1. An arc exists only from the level i+1 to the level i. There is only one gate at the level 1 whose output is the output of the circuit. The inputs of gates at the bottom level consist of literals x_i and $\neg x_i$. The gates at the same level are of the same type. Two gates at two adjacent levels are of different types. For simplicity, we assume in the following that the word circuit means the circuit of the type that we just described.

A circuit computes a Boolean function. For example, the circuit in Figure 1 computes $f(x_1,x_2)=(x_1 \vee \neg x_2) \wedge (\neg x_1 \vee x_2) \wedge (\neg x_1 \vee \neg x_2)$. The parity function p_n is defined as follows.

The work was supported in part by the NSF of China.

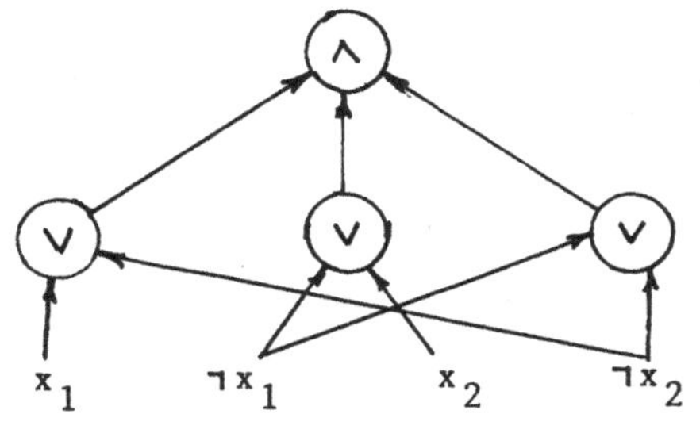

Figure 1

$$p_n(x_1,\ldots,x_n) = \begin{cases} 1, & \text{if } x_1+\cdots+x_n \text{ is odd,} \\ 0, & \text{if } x_1+\cdots+x_n \text{ is even.} \end{cases}$$

A circuit computing a parity function is called a parity circuit. The size of a circuit is defined to be the number of gates in the circuit. The depth of a circuit is the number of its levels. If we fix the depth d, then the minimum size of parity circuits is a function of the number of variables, $C_d(n)$.

The lower bound of $C_d(n)$ has been studied since 1961. Recently, it has been pushed up from $\Omega(n^{\log n})$ (Furst, Saxe and Sipser [2]) to $\Omega(\exp(n^{1/(4d)}))$ by Yao [5], to $\Omega(\exp((0.1)^{d/(d-1)}n^{1/(d-1)}))$ by Hastad[3] and to $\Omega(\exp(0.14(0.33n)^{1/(d-1)}))$ by Moran [4], where we denote $\exp(y)=2^y$. Furst, Saxe and Sipser [2] also translated an important open problem in structural complexity theory to the problem of proving an exponential lower bound for $C_d(n)$. Yao gave the first exponential lower bound and hence resolved the important open problem. Hastad [3], in addition, established the upper bound $O(n\cdot\exp(n^{1/(d-1)})$ for $C_d(n)$. There still exists a gap between the lower bound and the upper bound.

In this paper, we narrow the gap by presenting a new lower bound. The proof is based on an improved switching lemma. A new idea is introduced. It can also be used for improving other switching lemmas in [3,1]. We will include the works in the next paper.

2. Main Results

Consider a depth-2 circuit. It is called an AND-OR (OR-AND) circuit if the gate at the level one is V (Λ). The bottom fan-in of a circuit is the maximal number of fan-ins of gates at the bottom level. An AND-OR circuit is said to be able to switch to an OR-AND circuit with bottom fan-in s if there exists an OR-AND circuit with bottom fan-in s computing the function which the AND-OR circuit computes. The switching lemma tells us the possibility of such switching.

Consider a circuit G. A restriction ρ of G is a partial assignment to variables of G. $G!_\rho$ denotes the circuit resulting from G with the restriction ρ. To measure the possibility of the switching, each restriction is assigned with a probability. Here, we consider a probability space R_p with the property that a restriction ρ randomly chosen from R_p satisfies that

$$Pr(\rho(x_i)=0) = Pr(\rho(x_i)=1) = (1-p)/2$$

$$Pr(\rho(x_i)=x_i) = p$$

where p is a parameter.

Switching Lemma. Let G be an AND-OR circuit with bottom fan-in t and ρ a random restriction from R_p. Suppose

$$q = \frac{2p}{((1+p)^t+(1-p)^t)^{1/t}-(1-p)} \leqslant 1/2. \qquad (\$)$$

Then, the probability that $G!_\rho$ cannot be switched to an OR-AND circuit with bottom fan-in less than s is bounded by q^s.

Define deg(G) to be the degree of the polynomial obtained from G by simply expanding the function computed by G. For example, suppose that G computes $(x_1 \vee x_2) \wedge (x_1 \vee \neg x_2)$, then deg(G)=2 since $(x_1 \vee x_2) \wedge (x_1 \vee \neg x_2) = x_1 \vee x_1 \wedge \neg x_2 \vee x_1 \wedge x_2$ is of degree 2. If $\deg(G!_\rho) < s$ then it is clear that $G!_\rho$ can be switched to an OR-AND circuit with bottom fan-in less than s. Thus, to show Switching Lemma, it suffices to prove the following inequality

$$\Pr(\deg(G!_{\rho}) \geq s) \leq q^{s}.$$

The next lemma states a stronger result.

Lemma 1. Let G be an AND-OR circuit with bottom fan-in t and ρ a random restriction from R_p. Suppose that q satisfies ($). Then for any Boolean function F, we have

$$\Pr(\deg(G!_{\rho}) \geq s\colon F!_{\rho} \equiv 1) \leq q^{s}.$$

Proof. Write $G = \bigwedge_{i=1}^{w} G_i$. We prove the lemma by induction on w. For $w=0$, it is trivial since $G \equiv 1$. Next, consider $w \geq 1$. Note that

$$\Pr(\deg(G!_{\rho}) \geq s : F!_{\rho} \equiv 1)$$
$$= \Pr(\deg(G!_{\rho}) \geq s : F!_{\rho} \equiv 1 \wedge G_1!_{\rho} \equiv 1) \cdot \Pr(G_1!_{\rho} \equiv 1 : F!_{\rho} \equiv 1)$$
$$+ \Pr(\deg(G!_{\rho}) \geq s : F!_{\rho} \equiv 1 \wedge G_1!_{\rho} \not\equiv 1) \cdot \Pr(G_1!_{\rho} \not\equiv 1 : F!_{\rho} \equiv 1).$$

If $G_1!_{\rho} \equiv 1$, then $G!_{\rho} = (\bigwedge_{i=2}^{w} G_i)!_{\rho}$. Thus, by the induction hypothesis, we have

$$\Pr(\deg(G!_{\rho}) \geq s : F!_{\rho} \equiv 1 \wedge G_1!_{\rho} \equiv 1)$$
$$= \Pr(\deg(G!_{\rho}) \geq s : (F \wedge G_1)!_{\rho} \equiv 1) \leq q^{s}.$$

Therefore, the remainder is to show

$$\Pr(\deg(G!_{\rho}) \geq s : F!_{\rho} \equiv 1 \wedge G_1!_{\rho} \not\equiv 1) \leq q^{s}.$$

Let $G_1 = \bigvee_{j \in T} y_j$ where $y_j = x_j$ or $\neg x_j$. Let H_i be the clause obtained from G_i by deleting all literals belonging to $X = \{x_j, \neg x_j : j \in T\}$. Define $H = \bigwedge_{i=2}^{w} H_i$. It is an important observation that every term in the expansion of $G!_{\rho}$ is a product of some literals in X and a subterm in the expansion of $H!_{\rho}$, where a subterm is defined to be the product of some literals in a term. Let $\rho(J)=*$ denote the event that for any $j \in J$, $\rho(y_j)=y_j$ and let $\rho(J)=0$ denote the event that for every $j \in J$, $\rho(y_j)=0$. Define ρ' to be the restriction on the variables not appearing in G_1, which is induced by ρ. Then, we have

$$\Pr(\deg(G!_{\rho}) \geq s : F!_{\rho} \equiv 1 \wedge G_1!_{\rho} \not\equiv 1)$$
$$= \sum_{J \subseteq T,\ J \neq \emptyset} \Pr(\deg(G!_{\rho}) \geq s : F!_{\rho} \equiv 1 \wedge \rho(J)=* \wedge \rho(T \setminus J)=0)$$
$$\cdot \Pr(\rho(J)=* \wedge \rho(T \setminus J)=0 : F!_{\rho} \equiv 1 \wedge G_1!_{\rho} \not\equiv 1)$$
$$\leq \sum_{J \subseteq T,\ J \neq \emptyset} \Pr(\deg(H!_{\rho'}) \geq s - |J| : F!_{\rho} \equiv 1 \wedge \rho(J)=* \wedge \rho(T \setminus J)=0)$$

$$\cdot \Pr(\rho(J)=* \wedge \rho(T\setminus J)=0 \ :\ F!_\rho \equiv 1 \wedge G_1!_\rho \not\equiv 1)$$

$$= \sum_{J\subseteq T, J\neq\emptyset} \Pr(\deg(H!_{\rho'}) \geq s-|J| \ :\ F'!_{\rho'} \equiv 1)$$

$$\cdot \Pr(\rho(J)=* \wedge \rho(T\setminus J)=0 \ :\ F!_\rho \equiv 1 \wedge G_1!_\rho \not\equiv 1)$$

where F' is obtained from F by substituting $y_j=0$, $j \in T\setminus J$ into it. By the induction hypothesis, we have

$$\Pr(\deg(H!_{\rho'}) \geq s-|J| \ :\ F'!_{\rho'} \equiv 1) \leq q^{s-|J|}.$$

Thus,

$$\Pr(\deg(G!_\rho) \geq s \ :\ F!_\rho \equiv 1 \wedge G_1!_\rho \not\equiv 1)$$

$$\leq \sum_{J\subseteq T, J\neq\emptyset} q^{s-|J|} \Pr(\rho(J)=* \wedge \rho(T\setminus J)=0 \ :\ F!_\rho \equiv 1 \wedge G_1!_\rho \not\equiv 1).$$

We want to write the right hand side into the following form

$$\sum_{J\subseteq T} a_{|J|} \Pr(\rho(J)=* \ :\ F!_\rho \equiv 1 \wedge G_1!_\rho \not\equiv 1).$$

Since

$$\Pr(\rho(J)=* \ :\ F!_\rho \equiv 1 \wedge G_1!_\rho \not\equiv 1)$$

$$= \sum_{J'\supseteq J} \Pr(\rho(J')=* \wedge \rho(T\setminus J')=0 \ :\ F!_\rho \equiv 1 \wedge G_1!_\rho \not\equiv 1)$$

we have

$$\sum_{J\subseteq T} a_{|J|} \Pr(\rho(J)=* \ :\ F!_\rho \equiv 1 \wedge G_1!_\rho \not\equiv 1)$$

$$= \sum_{J'\subseteq T} \Pr(\rho(J')=* \wedge \rho(T\setminus J')=0 \ :\ F!_\rho \equiv 1 \wedge G_1!_\rho \not\equiv 1)$$

$$\cdot \sum_{J\subseteq J'} a_{|J|}.$$

Setting

$$q^{s-|J'|} = \sum_{J\subseteq J'} a_{|J|} = \sum_{k=0}^{|J'|} a_k \binom{|J'|}{k}, \quad 0=a_0,$$

by the inversion formula, we can obtain

$$a_k = ((1-q)^k - (-q)^k)\cdot q^{s-k}.$$

Therefore,

$$\Pr(\deg(G!_\rho) \geq s \ :\ F!_\rho \equiv 1 \wedge G_1!_\rho \not\equiv 1)$$

$$\leq \sum_{J\subseteq T} ((1-q)^{|J|} - (-q)^{|J|})\cdot q^{s-|J|} \Pr(\rho(J)=* : F!_\rho \equiv 1 \wedge G_1!_\rho \not\equiv 1).$$

Note that

$$\Pr(F!_\rho \equiv 1 \ :\ \rho(J)=* \wedge G_1!_\rho \not\equiv 1) \leq \Pr(F!_\rho \equiv 1 : G_1!_\rho \not\equiv 1)$$

since requiring some variables not to be assigned with value 0 or 1 cannot increase the probability that a function is determined. Thus, we have

$$
\begin{aligned}
& Pr(\rho(J)=* \; : \; F!_\rho \equiv 1 \wedge G_1!_\rho \not\equiv 1) \\
&= Pr(\rho(J)=* \wedge F! \equiv 1 \wedge G_1!_\rho \not\equiv 1)/Pr(F!_\rho \equiv 1 \wedge G_1!_\rho \not\equiv 1) \\
&= (Pr(F!_\rho \equiv 1 \; : \; \rho(J)=* \wedge G_1!_\rho \not\equiv 1)/Pr(F!_\rho \equiv 1 \; : \; G_1!_\rho \not\equiv 1)) \\
&\qquad \cdot Pr(\rho(J)=* : G_1!_\rho \not\equiv 1) \\
&\leq Pr(\rho(J)=* : G_1!_\rho \not\equiv 1) = (2p/(1+p))^{|J|}.
\end{aligned}
$$

Therefore

$$
\begin{aligned}
& Pr(\deg(G!_\rho) \geq s : F!_\rho \equiv 1 \wedge G_1!_\rho \not\equiv 1) \\
&\leq \sum_{J \subseteq T} ((1-q)^{|J|}-(-q)^{|J|}) \cdot q^{s-|J|}(2p/(1+p))^{|J|} \\
&= \sum_{k=0}^{|T|} \binom{|T|}{k}\left(\left(\frac{1-q}{q}\right)^k-(-1)^k\right) \cdot q^s(2p/(1+p))^k \\
&\leq q^s \sum_{k=0}^{t} \binom{t}{k}\left(\left(\frac{1-q}{q}\right)^k -(-1)^k\right)\left(\frac{2p}{1+p}\right)^k \\
&= q^s\left(\left(\frac{2p}{q(1+p)} + \frac{1-p}{1+p}\right)^t - \left(\frac{1-p}{1+p}\right)^t\right) = q^s. \qquad \square
\end{aligned}
$$

We will prove our lower bound by induction on the depth of circuits. The following lemma states a known fact which will be the base of the induction. For seeking the completeness, we also include a proof here.

Lemma 2. Every depth-2 parity circuit is of size $2^{n-1}+1$ and bottom fan-in n where n is the number of its variables.

Proof. We first consider an OR-AND circuit computing p_n, and claim that each AND-gate (at the bottom level) must have fan-in n. This means that every variable should appear in the inputs of the AND-gate. By the contradiction, suppose that the variable x_i does not appear in the inputs of the AND-gate. Then, there is an assignment which does not give a value to x_i but make the parity function p_n equal to 1. It is impossible since p_n should take different values as x_i takes different values. Now, there is a unique assignment for an AND-gate to get the value 1. Moreover, there are exactly 2^{n-1} truth-assignments for p_n.

Therefore, the parity circuit must have 2^{n-1} AND-gates and hence is of size $2^{n-1}+1$. For an AND-OR parity circuit, by DeMorgan' law, we can consider a corresponding OR-AND circuit computing p_n. □

Now, we are ready to show our main theorem.

Theorem 1. Let u and v be two positive numbers satisfying the following conditions

(1) $u < 0.5 \cdot \ln((1+\sqrt{5})/2)$, and

(2) $2^v(2u/\ln(e^{2u}+1))^u < 1$.

Then, for sufficiently large n, the parity function p_n can not be computed by a depth-d circuit containing at most $\exp(v \cdot n^{1/(d-1)})$ gates not at the bottom level and with bottom fan-in at most $u \cdot n^{1/(d-1)}$. The same result holds for the function $\neg p_n$.

Proof. We prove it by induction on d. For d=2, it follows from the Lemma 2. Next, we consider $d \geq 3$. By the contradiction, suppose that there exists a depth-d circuit C containing g ($\leq \exp(v \cdot n^{1/(d-1)})$) gates not at the bottom level and with bottom fan-in t ($\leq u \cdot n^{1/(d-1)}$) which computes either p_n or $\neg p_n$. Without loss of generality, assume that its bottom two levels are in the AND-OR form since, for otherwise, we can consider $\neg C$ instead of C. Let ρ be a random restriction from R_p where $p=n^{-1/(d-1)}$. There are at most g AND-OR circuits at the bottom two levels of $C\!\restriction_\rho$. Let A denote the event that all of such AND-OR circuit can be switched to OR-AND circuits with bottom fan-in less than $s = u \cdot n^{1/(d-1)}$. Let B denote the event that $C\!\restriction_\rho$ contains at least $pn = n^{(d-2)/(d-1)}$ remaining variables. Since pn is the expectation of the number of remaining variables, we have $\Pr(B) > 1/3$. By (1),

$$\lim_{n\to\infty} q = \lim_{n\to\infty} \frac{2p}{((1+p)^t+(1-p)^t)^{1/t}-(1-p)}$$

$$\leq \lim_{p\to 0} \frac{2p}{((1+p)^{u/p}+(1-p)^{u/p})^{p/u}-(1-p)}$$

$$= 2u/\ln(e^{2u}+1) < 1/2.$$

Thus, for sufficiently large n, $q \leq 1/2$. By Switching Lemma,

$$\Pr(A) \geq (1-q^s)^g \geq 1-g\cdot q^s \geq 1-(2^v q^u)^{n^{1/(d-1)}}$$

By (2), $\lim_{n\to\infty} 2^v q^u < 1$. Thus, $\lim_{n\to\infty} \Pr(A) = 1$ and hence $\Pr(A \wedge B) = \Pr(A)+\Pr(B)-\Pr(A \vee B) \geq \Pr(A)+\Pr(B)-1 \rightarrow 1/3$ as $n\to\infty$. This means that for sufficiently large n, there is a restriction ρ such that event A and event B both occur. Let G be the circuit obtained from $C|_\rho$ by switching all AND-OR circuits at the bottom two levels into OR-AND circuits with bottom fan-in less than s. Since the second bottom level and the third bottom level both consist of OR-gates, we can merge them into one level. Thus, G is actually a depth-(d-1) circuit satisfying the following conditions.

(a) It has m ($\geq pn$) variables.

(b) It contains $\leq g$ ($\leq \exp(v\cdot n^{1/(d-1)}) \leq \exp(v\cdot m^{1/(d-2)})$) gates not at the bottom level.

(c) Its bottom fan-in is less than $s = u\cdot n^{1/(d-1)}$ $u\cdot m^{1/(d-2)}$.

Moreover, $C|_\rho$ computes either p_m or $\neg p_m$, so does G, contradicting the induction hypothesis. □

<u>Theorem</u> 2. Let u and v be two positive number satisfying the conditions (1) and (2). Let w be a positive number satisfying

(3) $2^v(2w/(1+w))^u < 1$ and $w < 1/3$.

Then, for sufficiently large n, the parity function p_n can not be computed by a depth-d circuit of size not greater than $\exp(v(wn)^{1/(d-1)})$.

Proof. Consider a depth-d parity circuit C of size g. By the contradiction, suppose $g \leq \exp(v(wn)^{1/(d-1)})$. We treat C as a depth-(d+1) circuit with bottom fan-in t=1 and containing g gates not at the bottom level. Choose p=w and $s = u(wn)^{1/(d-1)}$. Applying the Switching Lemma, we can obtain a depth-d circuit G satisfying the following conditions.

(a) G computes either p_m or $\neg p_m$ where $m \geq pn = wn$.

(b) G contains at most $\exp(v \cdot m^{1/(d-1)})$ gates not at the bottom level.

(c) Its bottom fan-in is less than $s \leq u \cdot m^{1/(d-1)}$.

By Theorem 1, such G cannot exist if n is sufficiently large. □

Corollary. For sufficiently large n,

$$C_d(n) > \exp((0.24)^{d/(d-1)} n^{1/(d-1)}).$$

Proof. It is easy to verify that u=v=w=0.24 satisfy the conditions (1), (2) and (3) in Theorems 1 and 2. □

3. Discussion

The new idea which enable us to obtain the current lower bound is using the notion of deg(G). Comparing it with the notions of minterm(G) and AND(G) in [2,3,4,5], it is easy to see that minterm(G) $\geq$ deg(G) $\geq$ AND(G). The superiority of deg(G) is that it is easier to be dealed with. In fact, our proof for Switching Lemma is simpler and more understandable.

The condition ($) in Switching Lemma seems a little cumbersome.We conjecture that ($) can be deleted. In fact, it is easy th show the truth for t=1. This also implies that the condition $w < 1/3$ can be eliminated from Theorem 2.

Acknowledgments. The author wish to thank Professor Ker-I Ko for his corrections and helpful suggestions.

References

[1] W. Aiello, S. Goldwasser and J. Hastad, On the power of interaction, to appear.

[2] M. Furst, J. Saxe and M. Sipser, Parity, circuits, and the polynomial time hierarchy, Math. System Theory, 17 (1984),13-27.

[3] J. Hastad, Almost optimal lower bounds for small depth circuits, Proceedings of 18th Annual ACM Symposium on Theory of Computing, 1986, pp.6-12.
[4] S. Moran, An improvement of the Hastad-Yao lower bound on parity circuits, preprints.
[5] A.C. Yao, Separating the polynomial-time hierarchy by oracles, Proceedings 26th Annual IEEE Symposium on Foundations of Computer Science, 1985,pp.1-10.

HOW TO DESIGN ROUND ROBIN SCHEDULES

F. K. Hwang

AT&T Bell Laboratories
Murray Hill, NJ 07974

1. Introduction

We will call the basic unit of competition in a tournament a *team*. This team can be a genuine *team* as in basketball or football, or a pair of players as in doubles tennis or duplicate bridge, or even just a single player as in singles tennis or boxing. What is important is that the scheduling considers the team, regardless of how many players it has, as a single, indivisible unit. A *match* considered here involves two opposing teams unless specified otherwise. When the number of teams is not too large in a tournament, then the round robin schedule is often used for two reasons: (i) it provides the most balanced comparisons among the teams, (ii) for social games, it allows every team to meet every other team.

A *round robin schedule (RRS)* S_n with n teams is characterized by two conditions: (i) Every team opposes every other team exactly once. (ii) The $\binom{n}{2}$ matches in the schedule are partitioned into *rounds* such that every team plays exactly one match in each round.

Clearly, n must be even to satisfy condition (ii). We assume that this is the case throughout this paper unless specified otherwise. It is also clear from condition (ii) that an S_n has $n - 1$ rounds each of which consists of $n/2$ matches.

Let Z_k denote the set $\{0, 1, \ldots, k - 1\}$. Since we will use modular addition in deriving S_n, we will usually index the teams and rounds by members of $Z_{n-1} \cup \infty$ where ∞ denotes a team with special movement in the schedule. Many designs of S_n have been proposed. The most well known one [22], and perhaps also the simplest one will be denoted by F_n (called a *circle design* in [16]). An interesting geometric interpretation of F_n is to place team $n - 1$ at the center of a regular $(n - 1)$-gon and the other teams at the $n - 1$ corners in numerical order. Each edge connecting two teams represents a match. Matches in round i then consist of the edge $[n - 1, i]$ and $n/2 - 1$ other edges perpendicular to it. This construction also makes clear the relation between S_n and a 1-factorization of the complete graph with n vertices. Figure 1 gives E_6.

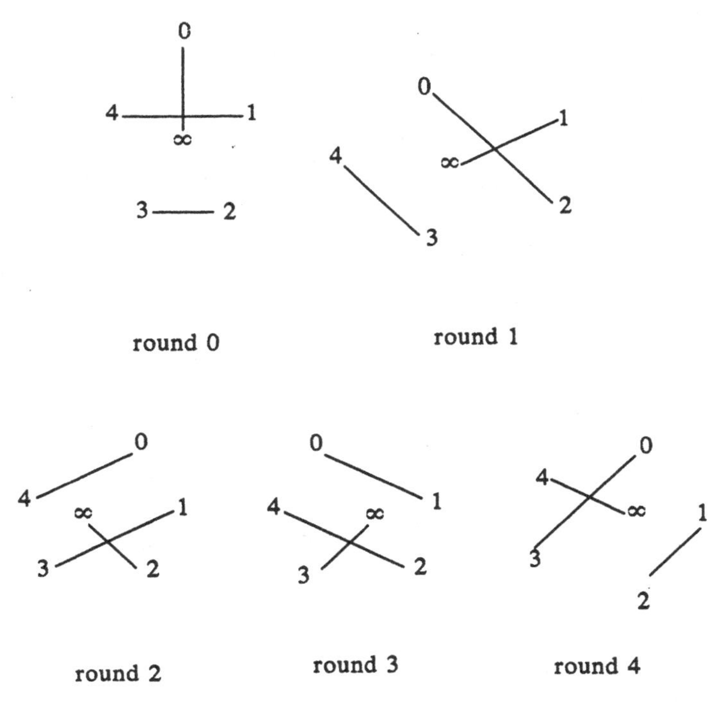

FIGURE 1. *THE SCHEDULE* F_6

Define $Z_k^* = Z_k - \{0\}$. A *starter* of Z_{n-1} is a partition of Z_{n-1}^* into $n/2 - 1$ pairs (z_i, z_i') such that the set of symmetric differences $\{\pm(z_i - z_i') : i = 1, \ldots, n/2 - 1\} = Z_{n-1}^*$. Let (i, j) denote a match between teams i and j. Then a cyclic S_n on teams $Z_{n-1} \cup \infty$ can be obtained from a starter of Z_{n-1} by setting round 0 to be $\{(\infty, 0), (z_1, z_1'), \ldots, (z_{n/2-1}, z_{n/2-1}')\}$, and "cyclically developing" round 0. Namely, round $r + 1$ is obtained from round r by substituting team $i + 1$ for team $i (\mathrm{mod}\ n - 1)$ and letting team ∞ stay put for $r = 0, 1, \ldots, n - 2$. Let *column k* consist of the pair (z_k, z_k') (column 0 starts with the pair $(\infty, 0)$) and its subsequent substitutions. Note that the differences $\pm(a_k - b_k)$ are preserved in column k and each half of column k runs through the set Z_{n-1} for $k = 1, \ldots, n/2 - 1$. Hence two teams i and j, neither being ∞, with $i - j = a_k - b_k$ or $b_k - a_k$ must oppose each other in column k, while team ∞ oppose every other team in column 0. F_n can then be described by the starter $\{(n - i - 1, i) : i = 1, \ldots, n/2 - 1\}$. Note that a cyclic development of a starter yields a *cyclic* S_n which simplifies the movement of teams from round to round, i.e., team ∞ stays put throughout and team $i \neq \infty$ needs only to follow the movement of team $i - 1$. Figure 2 gives F_6 in three columns where each row is a round.

$(\infty, 0)$	(4,1)	(3,2)
$(\infty, 1)$	(0,2)	(4,3)
$(\infty, 2)$	(1,3)	(0,4)
$(\infty, 3)$	(2,4)	(1,0)
$(\infty, 4)$	(3,0)	(2,1)

FIGURE 2. *THE SCHEDULE F_6 IN COLUMNS*

Although designing S_n is easy and well known, the job becomes harder when other factors are considered. We will introduce these factors in the following sections. In general, we will give the constructions, if simple enough, but not the proofs. For more complicated constructions we will simply mention them and then refer the reader to the most relevant literature.

2. The Balance of Carry-Over Effects

We note that in F_6 teams 2,4,1 before they play team 0 (in rounds 2,3,4 respectively), all play team 3 in the immediately preceding rounds. If team 3 is a strong team which crushes teams 2,4,1, then these teams could become demoralizing in the succeeding rounds which will benefit team 0. Roselle [27] introduced the concept of balancing the carry-over effects over the rounds. More specifically, suppose that team i plays against team i_r in round r. Let j_{r-1} be the opponent of i_r in round $r - 1$. Then we want $J_i = \{j_{r-1} : r = 1, \ldots, n - 2\}$ to contain every team except i once. However, this is impossible since J_i has only $n - 2$ members while there are $n - 1$ teams not counting i. This is remedied by adding j_{n-2} to J_i where j_{n-2} is the opponent of i_0 in round $n - 2$. This addition can be justified if the same RRS is run continuously (repeating itself). Roselle showed that such a schedule can be constructed if n is a power of 2, and conjectured that this condition is also necessary. We now give a modified version E_n to Roselle's construction such that it is simpler in concept, it yields a cyclic schedule (Roselle's does not) and it unifies with some later construction.

For $n = 2^m$ let x be a primitive root of $GF(n) = PG(m, 2)$. Index round r by x_r for $r = 0, 1, \ldots, n - 2$. At round 0 team i opposes team $i + 1$. The other rounds are obtained from round 0 by a cyclic development with team 0 staying put. It is easily verified that at round r team i opposes team $i + x^r$ for $r = 0, 1, \ldots, n - 2$. Hence E_n is a RRS. Since $x^i + i^r = x^r$ and $i^r + j^{r-1}$ we have $j^{r-1} = x^{r-1} - x^r + x^i$ which are distinct over r for each fixed i. Figure 3 gives R_8 with $x^3 = x^2 + 1$, $x^4 = x^2 + x + 1$, $x^5 = x + 1$, $x^6 = x^2 + x$ and $x^7 = 1$.

$(0, 1)$	$(x, x + 1)$	$(x^2, x^2 + 1)$	$(x^2 + x + 1, x^2 + x)$
$(0, x)$	$(x^2, x^2 + x)$	$(x^2 + 1, x^2 + x + 1)$	$(x + 1, 1)$
$(0, x^2)$	$(x^2 + 1, 1)$	$(x^2 + x + 1, x + 1)$	$(x^2 + x, x)$
$(0, x^2 + 1)$	$(x^2 + x + 1, x)$	$(x + 1, x^2 + x)$	$(1, x^2)$
$(0, x^2 + x + 1)$	$(x + 1, x^2)$	$(x^2 + x, 1)$	$(x^2, x^2 + 1)$
$(0, x + 1)$	$(x^2 + x, x^2 + 1)$	$(1, x)$	$(x^2, x^2 + x + 1)$
$(0, x^2 + x)$	$(1, x^2 + x + 1)$	(x, x^2)	$(x^2 + 1, x + 1)$

FIGURE 3. *THE SCHEDULE* E_8

Roselle [27] conjectured that $n = 2^m$ is also necessary for E_n to exist.

3. The Balance of Courts

We assume that a match is played on a *court* and a court can be used only for one match at a given round. Suppose that the courts are of different qualities. Then it is desirable that the courts are distributed evenly to the teams. Can we assign matches to courts such that no team will play more than two matches at any court? (This implies that each team plays two matches at each court except only one match is played at one court.) Such a design T_n was first studied by Gelling and Odeh [14] and was called a *balanced tournament design* in [31].

For $n \not\equiv 1 \pmod 3$ Haselgrove and Leech [16] reported Gray's construction which simply interchanges the match at site 0 with a match at some other site at every round except one from F_n. The other site starts at site 1, and then hops by 2 for each round except it hops by one at either the last or the next to last site. Whenever site 0 or site $n/2$ is reached, the hop reverses its direction. It is easily verified that the hop stays at site 0 at round $n/2 - 1$, so no interchange takes place. Figure 4 gives a T_{10} obtained from F_{10} (the bracketed matches are interchanged).

Site / Round	0	1	2	3	4
0	[8,1]	[∞, 0]	(7,2)	(6,3)	(5,4)
1	[7,4]	(0,2)	(8,3)	[∞, 1]	(6,5)
2	[7,6]	(1,3)	(0,4)	(8,5)	[∞, 2]
3	[1,5]	(2,4)	[∞, 3]	(0,6)	(8,7)
4	(∞, 4)	(3,5)	(2,6)	(1,7)	(0,8)
5	[3,7]	(4,6)	[∞, 5]	(2,8)	(1,0)
6	[2,1]	(5,7)	(4,8)	(3,0)	[∞, 6]
7	[4,1]	(6,8)	(5,0)	[∞, 7]	(3,2)
8	[7,0]	[∞, 8]	(6,1)	(5,2)	(4,3)

FIGURE 4. *A SCHEDULE* T_{10}

The reader is referred to Schellenberg, Rees and Vanstone [31] for a general construction of T_n for all values of $n \neq 4$.

4. The Balance of Boards

In a bridge game a court is simply a table where two teams can sit around to play a board. Since most tables are of uniform quality and do not affect the match in any discernible way, there is little motivation to balance the tables. However, the role of courts is taken over by the role of boards. Although the use of boards in bridge is comparable to the use of balls in tennis, the concept of a board is unique in the sense that no team can play the same board twice (this is because the bridge hands are unfolded during the play and estimating the unseen hands is an important part of the game). Assume that one board is played at each match. Clearly, a minimum of $n - 1$ boards, which will be indexed by elements of Z_{n-1}, are needed for S_n. Furthermore, although a board can be duplicated, it takes time as well as additional equipment to do so. Therefore, it is desirable to assign a given board only to one match at a round. An S_n satisfying the board requirements can be laid out as a $(n - 1) \times (n - 1)$ square where rows represent rounds and columns represent boards. Each cell contains either a pair of teams (i, j) or is empty. Cell (r, c) containing the pair (i, j) is interpreted as that team i and team j play board c against each other at round r. The requirements are that every team must appear exactly once in each row and each column. Such a square is called a *Room square* of side $n - 1$ and was first studied by Room [28]. We will denote it by R_n.

Stanton and Mullin [33] introduced a systematic way to construct R_n. An *adder* of a starter $\{(z_i, z_i')\}$ of Z_{n-1} is a set of integers $\{a_1, \ldots, a_{n/2-1}\}$ from Z_{n-1}^* such that $a_i + z_i$, $a_i + z_i' : i = 1, \ldots, n/2 - 1$ are all distinct elements in Z_{n-1}^*. It was shown in [33] that an R_n can be constructed from a starter of Z_{n-1} and its adder by assigning pair $(\infty, 0)$ to play board 0 and pair (z_i, z_i') to play board $-a_i$ in round 0, and a cyclic development (running diagonally), which includes substituting board $b + 1$ for board b (mod $n - 1$), to obtain the other $n - 2$ rounds. A starter is called *strong* if $z_i + z_i'$, $i = 1, \ldots, n/2 - 1$ are all distinct elements of Z_{n-1}^*. Mullin and Nemeth [24] showed that if a starter is strong, then we can use

$\{-(z_i + z_i')\}$ to be the adder. They also proved that for $n - 1 \equiv 3 \pmod 4$ a prime power and x a primitive root, then $\{(x^{2i-2}, x^{2i-1}) : i = 1, \ldots, n/2 - 1\}$ is a strong starter. Figure 5 gives an R_8 (where rows are rounds and columns are boards) constructed by this method ($x = 3$).

(∞,0)	(2,6)	(4,5)		(1,3)		
	(∞, 1)	(3,0)	(5,6)		(2,4)	
		(∞, 2)	(4,1)	(6,0)		(3,5)
(4,6)			(∞, 3)	(5,2)	(0,1)	
	(5,0)			(∞, 4)	(6,3)	(1,2)
(2,3)		(6,1)			(∞, 5)	(0,4)
(1,5)	(3,4)		(0,2)			(∞, 6)

FIGURE 5. *A SCHEDULE* R_8

Based on this "starter and adder" construction, along with some multiplication theorem to propagate the squares and some special constructions to fill the gaps, it is finally proved that R_n exists for every even n except 4 and 6. The reader is referred to [25] for a condensed proof.

5. The Balance of Competitors

There is another peculiarity of bridge due to the use of a board. A board consists of four randomly dealt hands with one team holding two of them, usually designated as the NS hands, and the other team holding the EW hands. The outcome of a board is determined by the relative skills of the two opposing teams as well as by the relative strengths of the NS hands versus the EW hands. Therefore, to achieve any accuracy in evaluating the teams, we must eliminate the factor of board. This is done by comparing the scores of all teams holding the NS hands on a given board and rank them. Since each board is a zero-sum game for the two opposing teams, the ranks of the teams holding the EW hands are also determined. For a given board the teams which hold the same NS (or EW) hands are said to *compete* against each other. A S_n is called a *balanced Howell* rotation, and denoted by B_n, if it satisfies the additional condition that every team competes against every other team equally often (in fact, $n/2 - 1$ times).

A B_n is called a *complete balanced Howell rotation*, or a *balanced room square*, and denoted by C_n, if it is also a Room square, i.e., every board can be assigned to at most one match in a round. On the other hand, a Room square is balanced if its starter can be given as ordered pairs (z_i, z_i'), where $\{z_i : i = 1, \ldots, n/2 - 1\}$ and $\{z_i' : i = 1, \ldots, n/2 - 1\}$ are complementary difference sets, i.e., the symmetric differences generated by these two sets run through Z_{n-1}^*. We will call such a starter a *balanced starter*. Note that B_n can be constructed from a balanced starter since we can always assign board r to all matches at round r for $r \in Z_{n-1}^*$.

B_n and C_n have been used in bridge tournaments [7] for a long time. Since their construction is quite difficult, the "competition" condition is often compromised in practice. Parker and Mood [26] first brought these schedules to the attention of mathematicians and explicitly stated that condition. They proved that a necessary condition for B_n, hence C_n, to exist is that $n \equiv 0 \pmod 4$ and they constructed C_n for n up to 16. Berlekamp and Hwang [6] gave the first systematic construction of C_n. Interpreting in the language of "starters and adders", it was shown in [6] that for

$n - 1 \equiv 3 \pmod 4$ a prime power and x a primitive root, $\{(-x^{2i}, x^{2i}) : 0 \le i \le n/2 - 2\}$ is a balanced starter with adder $\{-x^{2i}/y : 0 \le i \le n/2 - 2\}$ where $y^2 - 1$ is a quadratic residue (the existence of such a y was proved). Later, Schellenberg [30] pointed out that the strong starter of Mullin and Nemeth, if viewed as ordered pairs, is also balanced.

Schellenberg [30] also gave a composition method which combines a pair of C_n into C_{2n} under certain conditions. In particular, he showed that the method works when $n - 1 \equiv 3 \pmod 4$ is a prime power. Hwang, Kang and Yu [21] gave a "starter" version of Schellenberg's conditions. A balanced starter $\{(z_i, z_i')\}$ of Z_{n-1} is called *skew* if $\pm(z_i + z_i')$ run through Z_{n-1}^*, and is called *symmetric* if $\{z_i\} = \{-z_i\}$. It was shown in [21] that the existence of a symmetric skew balanced starters (*SSBS*) for $Z_{n/2-1}$ implies the existence of C_n. Let x be a primitive root of $GF(n/2 - 1)$. For $u \in GF(n/2 - 1)$ equal x^v we define $\log_x u = v$. Define $M_t = \{x^{2^m i + t} : 0 \le i < k\}$. Recently, Du and Hwang [13], culminating a series of research by various authors, proved that if $n/2 - 1 \equiv 2^m k + 1$ is a prime power where $m \ge 2$ and $k \ge 3$ is odd, i.e., the prime power is not of the Fermat type, then

$$\{(yu, y) : y \in M_{2j+2},\ 0 \le j < 2^{m-2}\} \cup \{(yv, y) : y \in M_{2^{m-1}+2j+2},\ 0 \le j < 2^{m-}$$

is a SSBS, where u and v (whose existences were proved) are two elements of M_t for some odd t which satisfy the following three conditions:

(i) $\log_x(u - 1) \equiv \log_x(v - 1) + 1 \pmod 2$

(ii) $\log_x(u + 1) \equiv \log_x(v + 1) + 1 \pmod 2$

(iii) u is a primitive root.

It was conjectured in [20] that no SSBS exists for $k = 1$. More is known about the construction of B_n. Hanner [15] showed that a balanced starter of Z_{n-1} hence B_n, always exists when $n - 1 \equiv 2^m k + 1$ is an odd prime power at least 7. For $k = 1$ let x be a primitive root such that $x^3 + x^2 + x = t^4$ for some $t \in Z_{n-1}^*$ (the existence for such an x was proved). Then the starter consists of the pairs:

$$(x^{4i}, x^{4i+1}),\ (x^{4i+2}, x^{4i+3}),\ (-x^{4i+1}, -x^{4i+2}),\ (-x^{4i}, -x^{4i+3}) \text{ for } 0 \le i < 2^m$$

For $k > 1$ let x be any primitive root and define M_t as before. Then Hanner showed that

$$\{(y, uy) : y \in M_{2j},\ 0 \le j < 2^{m-1}\} \cup \{(y, vy) : y \in M_{2^{m-2}+2j},\ 0 \le j < 2^{m-2}\}$$

is a balanced starter where u and v are two elements in M_t for some odd t with one of them a quadratic residue and the other a nonresidue.

Hwang [19] showed that a B_{2n} can be constructed from a B_n. Since B_2 exists, this implies B_n exists for all $n = 2^m$. Du and Hwang [12] gave a more powerful multiplication theorem. A starter $\{(z_i, z_i')\}$ of Z_{n-1} is called *partitionable* if the pairs can be partitioned into two sets, P_1 with $n/4$ pairs and P_2 with $n/4 - 1$ pairs, such that P_1 and $P_2 \cup \{0\}$ are complementary difference sets. It was shown in [12] that a B_{mn} can be constructed from a B_m, a B_n and a partitionable starter of Z_{m-1}. The construction of partitionable starters was also studied in [20].

For $n = 2^m$ Anderson [1] showed that the starter of E_n (given in Section 2) can be ordered into a balanced starter by choosing a hyperplane H in $PG(m - 1, 2)$ such that $1 \notin H$, and by assigning the element in H as the first element in each pair of the starters. He also proved that if m is odd, then this procedure yields a balanced strong starter, hence C_n exists.

Finally, Du and Hwang [11] used Galois domain to give a construction of B_n when n is a product of two prime powers $P = p^r$ and $Q = q^s$ with $Q - P = 2$ and $q \neq 3$.

6. The Balance of the Home-and-Away Pattern

In most professional sports a match is usually played at the home town of one of the two opposing teams. It is generally recognized that the team which plays before its home town audience enjoys a home-court advantage. Thus we want to balance the number of home games and that of away games. Of course, for n even, a team plays an odd number of games and cannot achieve complete balance between these two numbers. The best one can do is to insist that they differ by exactly one. Besides balancing the numbers, it is also desirable to alternate the home and away game over the rounds to maximize the box office potential and also to avoid the travel fatigue imposed on the team from consecutive away games. de Werra [37] called a pair of consecutive home games or consecutive away games a *break*. He proved that a RRS must contain a minimum of $n - 2$ breaks by the following clever argument. Let S_i denote the home-and-away pattern sequence of team i, where the sequence runs over the $n - 1$ rounds. Then $S_i \neq S_j$ for $i \neq j$ since team i must oppose team j, one having a home game and the other an away game, at some round. But there are only two distinct sequences containing no break, i.e., one is $(H, A, H, \ldots, H)$ and the other $(A, H, A, \ldots, A)$. So at least $n - 2$ teams will have one break each. de Werra also gave a construction A_n which has only $n - 2$ breaks. Take the schedule F_n. Reverse the ordering of every other pair (starting from the first) in column 0 and reverse all pairs in columns with even index. For every match interpret the first team of the pair as the home team. Then teams 0 and $2n - 1$ have no break while every other team has one break. Figure 6 gives A_6.

$(\infty,0)$	$(4,1)$	$(2,3)$
$(1,\infty)$	$(0,2)$	$(3,4)$
$(\infty,2)$	$(1,3)$	$(4,0)$
$(3,\infty)$	$(2,4)$	$(0,1)$
$(\infty,4)$	$(3,0)$	$(1,2)$

FIGURE 6. *A SCHEDULE* A_6

In some real situations it is desirable to balance the home-and-away pattern between every two teams. Namely, two games will be played between every two teams with each team having home games once. This can be done by running A_n twice while reversing all pairs in the second run.

7. Odd Number of Teams

When n is odd, then one team must sit out (plays a *bye*) at a round. Hence the minimum number of rounds of an S_n is now n and condition (ii) is changed to:

(ii′) Every team plays at every round except one.

In general, if S_{n+1} exists, then S_n for odd n can be obtained from S_{n+1} by deleting all matches involving the team with the largest index. We will examine the effects of such deletions, if any, on the various properties.

(i) The properties of F_n and R_n are clearly preserved.

(ii) Schellenberg, van Rees and Vanstone [31] pointed out that, by deleting all games involving team ∞ in all F_{n+1}, T_n for odd n has every team play exactly twice on every court.

(iii) Berlekamp and Hwang [6] proved that the properties of C_n are preserved for $n \equiv 3 \pmod 4$ a prime power.

(iv) de Werra [37] proved that A_n has no break.

The two exceptions are

(v) For $n \equiv 1 \pmod 4$ Hwang [18] pointed out $B_n(C_n)$ can be constructed from a balanced (strong) starter of Z_n even though $B_{n+1}(C_{n+1})$ does not exist. When n is a prime power not of the Fermat type, a B_n was given by Hanner [15] and a C_n by Du and Hwang [13] as reported in Section 5.

(vi) For $n = 2^m - 1$ let E_n be obtained from E_{n+1} by deleting all matches involving team 0. The set J_i in E_{n+1} loses two members, one being 0 and the other being j_{i-1} (since team i has no opponent in round i). This is the best one can do to balance the carry-over effect when n is odd.

8. The Doubles

Let D_n denote a schedule satisfying the following two conditions (called a *balanced doubles schedule* in [17]):

(i) Each match consists of four teams divided into two opposing partnerships with two teams each.

(ii) Every team partners with every other team once and opposes every other team twice.

Lewis Carroll [10] seems to be the first one to study D_n for tennis tournaments. Later, Scheid [29] brought the problem to the attention of mathematicians. Yalavigi [39] gave a solution when $n = 4t + 1$ is a prime power. Bose and Cameron [8] laid down the fundamental methodology for constructing D_n and Healy extended it to complete the study. Let $D(K, v)$ denote a *pairwise balanced design* with blocks of sizes $k \in K$ of a v-set and every pair of elements appear together in exactly one block. Suppose that D_k exists for all $k \in K$. Then we can construct a D_v from $D(K, v)$ by replacing each of its block of size k by a D_k.

Suppose that a resolvable $D(K, v)$ with b blocks and D_{k+m}, $k \in K$, exist. Then we can construct a D_{v+m} for $m = 1$ or $4 \leq m \leq b$ by first adding a new block consisting of the m new elements if $m \geq 4$, then adding the jth new element to the blocks of the jth resolution for $j = 1, \ldots, m$, and finally, replacing each block of size y by a D_y using the same set of teams.

By applying the above methods on the resolvable $D(\{4\}, 12t + 4)$ whose existence for all t was shown by Hanani and using a few special constructions, it was shown [17] that D_n exists for all $n \geq 4$.

A D_n is called a *whist tournament*, denoted by W_n, by Moore [23] if it also satisfies the round condition:

(iii) Every team plays one match in every round.

Since a match requires four teams, n must be divisible by four for W_n. By drawing upon a rich repertoire of combinatorial designs, in particular, group divisible designs, Moore [23] and Baker [5] proved the existence of W_n for all $n \equiv 0 \pmod 4$ except possibly, $n = 132$, 152 or 264.

The questions of board-balanced and competitor-balanced have not been seriously studied for W_n.

9. Group Divisible Round Robin Schedules

A group divisible round robin schedule (GDRRS) means a partition of the n teams into groups of equal size such that only teams not in the same group oppose each other (once). When there are only two groups, the schedule is called *mixed* and the two groups *man* and *woman*. When the group size is two, the schedule is called *spouse-avoiding* (SA) and each group a *couple*. These concepts have been introduced for doubles except in a *mixed doubles* a team does not partner (instead of "not oppose") with any team of the same sex. We borrow them here for singles.

Both SA and mixed RRS are defined only for even number of teams. So we assume that $n = 2m$ in either schedule. An SARRS can be obtained from a corresponding RRS by deleting the last round and use the matches of that round to define the couples, i.e., team i and team j will form a couple if they play a match in the last round. Of course, the deletion of the last round will hurt the property of being balanced somewhat. But it is usually as balanced as any $(2m - 2)$-round schedule can be. The one property that we need to look deeper into is the board balance since it is conceivable that a board can be deleted along with a round so that every team still plays every board exactly once. Such a schedule is called a *Howell design* of side $2m - 2$, denoted by H_{2m}, which differs from R_{2m} by being one size smaller as a square and by containing m pairs less in the cells. Anderson [2,3], Schellenberg and Vanstone [32] proved that H_{2m} exists for all $m > 2$. *Balanced Howell designs* in the sense of balanced Room squares have not been studied.

A mixed S_{2m} has m rounds and can be represented by a Latin square of order m with rows as men, columns as women, and cell (i, j) containing the index of rounds where man i opposes woman j. A Latin square of order n is of course well known to exist for all n. A simple construction is to assign round $i + j \pmod m$ to cell (i, j).

With the Latin square interpretation it is easy to see that a mixed T_{2m} (which is also a Howell design of side m) exists if and only if a pair of orthogonal Latin square of order m exists, which means $m \neq 2$ or 6. If we introduce board as another factor independent of court, then we can achieve both court and board balance with 2^m teams if a set of three pairwise orthogonal Latin squares of order m exists. We can consider the balance of more factors in a similar way. It should be pointed out that if m is a prime power, then there exists a set of $m - 1$ orthogonal Latin squares.

If we want to balance the carry-over effect for just one group, say the man, then we can interchange the roles of men and rounds. Now a mixed E_{2m} is equivalent to a *column-complete* Latin square (every ordered pair (i, j) appears exactly once in the columns) of order m except in the latter case there is no carry-over effects from the last row to the first row. For m even Williams [38] gave the simple construction obtained by a cyclic development (mod $2m$) from the column whose transpose is: 0, $m - 1$, 1, $- 2$, ... , $m/2 - 1, m/2$. Figure 7 gives a column-complete Latin square of order 6.

0	1	2	3	4	5
5	0	1	2	3	4
1	2	3	4	5	0
4	5	0	1	2	3
2	3	4	5	0	1
3	4	5	0	1	2

FIGURE 7. *A COLUMN-COMPLETE LATIN SQUARE OF ORDER 6*

Note that woman 0 inherits the opponents of women 5,2,3,4,1 in rounds 1,2,3,4,5, respectively. There has been no study on mixed RRS which balance the carry-over effects for both men and women. Such a schedule is not equivalent to a Latin square with both row and column complete.

It is rather easy to construct mixed A_n. Again, take a Latin square with rows as men and columns as women, assign the game in cell (i, j) as a home game for $i\,(j)$ if $i + j$ (mod m) is odd (even). It is easily verified that there is no break for any man or woman.

We can also obtain mixed SARRS by deleting a round of the corresponding mixed RRS and using the pairs in that round to define couples.

10. Spouse-Avoiding Mixed Doubles

In a *spouse-avoiding mixed doubles schedule* two partnerships oppose each other in a match where a partnership always consists of a man and a woman. The requirements are that a team does not partner with its spouse, but partners with every team of the other sex once; and that a team does not oppose its spouse, but opposes every other team once. Note that the minimum number of rounds needed to satisfy either requirement is $m - 1$ (we will have two different numbers if spouse is not avoided). We will denote such a schedule by M_{2m}.

Brayton, Coppersmith and Hoffman [9] first studied M_{2m} and showed that the existence of M_{2m} is equivalent to the existence of a self orthogonal Latin square of order m. a Latin square is *self-orthogonal* if it is orthogonal to its transpose. The equivalence is seen by indexing both rows and columns by men, assigning to cell (i, j) (cell (j, i)) the woman partner of man i (j) in his match opposing man $j(i)$, and assigning to cell (i, i) the number i. It was shown in [9] that self orthogonal Latin squares of order M exists if and only if $m \neq 2, 3, 6$.

Willis [35], Wang and Wilson [36] studied M_{2m} which can be partitioned into rounds. The former showed their existence except for finitely many odd m; the latter proved the same for even m. Figure 8 gives a M_{2m} with $m = 8$ and seven rounds (both men and women are indexed by the set $\{0, 1, \ldots, 7\}$, with man following woman in each pair).

(1,4) vs (2,7)	(3,1) vs (4,5)	(5,3) vs (6,8)	(7,6) vs (8,2)
(1,7) vs (3,6)	(2,5) vs (7,3)	(4,2) vs (5,8)	(6,1) vs (8,4)
(1,6) vs (4,8)	(2,4) vs (8,5)	(3,7) vs (5,2)	(6,3) vs (7,1)
(1,3) vs (5,4)	(2,8) vs (6,7)	(3,5) vs (8,6)	(4,1) vs (7,2)
(1,5) vs (6,2)	(2,1) vs (3,8)	(4,3) vs (8,7)	(5,6) vs (7,4)
(1,8) vs (7,5)	(2,3) vs (4,6)	(3,2) vs (6,4)	(5,7) vs (8,1)
(1,2) vs (8,3)	(2,6) vs (5,1)	(3,4) vs (7,8)	(4,7) vs (6,5)

FIGURE 8. *A SCHEDULE* M_{16}

11. Conclusions.

Three basic methods set the tone for the design of RRS. The first is the circle design F_n, which, with modifications, is used in the construction of S_n, T_n and A_n for general n. The second is the starter-adder method used in the construction of R_n, B_n and C_n when either $n - 1$ or $n/2 - 1$ is an odd prime power. The third is using hyperplanes of $PG(m, 2)$ in the construction of E_n, B_n and C_n when $n = 2^m$.

For those varieties of RRS, D_n is constructed by using pairwise balanced designs, resolvable and group divisible. Mixed schedules are usually associated with Latin squares, orthogonal and self orthogonal, while SA schedules are typically obtained by deleting a round from the regular schedules except for H_n.

Both H_n and R_n are special cases of the Howell design of side s (with n teams), $s < n$, usually denoted by $H(s, n)$. $H(s, n)$ can be interpreted as a truncated Room square, with s rounds and s boards, such that each team oppose every other team *at most* once. The existence of $H(s, n)$ for all odd s, except $H(3,4)$, $H(5,6)$ and $H(5,8)$, was shown by Stinson [34]. The existence of $H(s, n)$ for all even s except $H(2,4)$ was shown by Andersen, Schellenberg and Stinson [4].

REFERENCES

[1] Andersen, B. A., Hyperplanes and balanced Howell rotations, Ars Combinatoria 13 (1983), 163-168.

[2] Andersen, B. A., Hyperovals and Howell designs, Ars Combinatoria 9 (1980), 29-38.

[3] Andersen, B. A. and Leonard, P. A., Sequencing and Howell designs, Pac. J. Math. 92 (1981), 249-256.

[4] Andersen, B. A., Schellenberg, P., and Stinson, D. R., J. Combin. Thy. Ser. A 36 (1984), 23-55.

[5] Baker, R., Whist tournament, Proc. 6th S-E Conf. Combin., Graph Thy., Comput. (1980), 89-100.

[6] Berelekamp, E. R., and Hwang, F. K., Contributions for balanced Howell rotations for bridge tournaments, J. Combin. Thy. 12 (1972), 159-166.

[7] Beynon, G. W., Duplicate Bridge Direction, Stuyvesant House, New York, (1944).

[8] Bose, R. C. and Cameron, J. M., The bridge tournament problem and calibration designs for comparing pairs of objects, J. Res. Nat. Bureau Standard B, Math. and Math. Phy. 69B (1965), 323-332.

[9] Brayton, R. K., Coppersmith, D., and Hoffman, A. J., Self-orthogonal Latin squares, in Atti Coll. Int. Teorie Combinatoire, Roma (1973), B. Segre Ed., Acad. Naz. Lincei, Rome 3, 509-517.

[10] Carroll, L., On the scheduling of lawn tennis tournaments, Complete works of Lewis Carroll, Modern Library.

[11] Du, D. Z. and Hwang, F. K., Balanced Howell rotations of the twin prime power type, Trans. Amer. Math. Soc. 271 (1982), 396-400.

[12] Du, D. Z. and Hwang, F. K., A multiplication theorem for balanced Howell rotations, J. Combin. Thy. Ser. A 37 (1984), 121-126.

[13] Du, D. Z. and Hwang, F. K., On the existence of symmetric skew balanced starters for odd prime powers, Proc. Amer. Math. Soc., to appear.

[14] Gelling, E. N. and Odeh, R. E., On 1-factorizations of the complete graph and the relation to round robin schedules, Proc. 3rd Manitoba Conf. on Numerical Math., Winnipeg, (1973), 213-221.

[15] Hanner, O., Construction of balanced Howell rotations for $2(p^r + 1)$ partnerships, J. Comb. Thy. Ser. A 33 (1982), 205-212.

[16] Haselgrove, J. and Leech, J., A tournament design problem, Amer. Math. Mon. 8 (1977), 198-201.

[17] Healy, P., Construction of balanced doubles schedules, J. Combin. Thy. Ser. A 29 (1980), 280-286.

[18] Hwang, F. K., Some more contributions on constructing balanced Howell rotations, Proc. 2nd Chapel Hill Conf. Combin. Math. Appl., (1970), 307-323.

[19] Hwang, F. K., New constructions for balanced Howell rotations, J. Combin. Thy. Ser. A 21 (1976), 44-51.

[20] Hwang, F. K., Strong starters, balanced starters and partitionable starters, Bull. Inst. Math., Acad. Sinica 11 (1983), 561-572.

[21] Hwang, F. K., Kang, C. D., and Yu, J. E., Complete balanced Howell rotations for $16k + 12$ partnerships, J. Combin. Thy. Ser. A 21 (1984), 66-72.

[22] Lockwood, E. H., American tournament, Note 1213, Math. Gaz. 20 (1936), 333.

[23] Moore, E. H., Tactical Memorandum I-III, Amer. J. Math. 18 (1956), 265-303.

[24] Mullin, R. C. and Nemeth, E., An existence theorem for Room squares, Canad. Math. Bull. 12 (1969), 493-497.

[25] Mullin, R. C. and Wallis, W. D., The existence of Room squares, Aequationes Math. 13 (1975), 1-7.

[26] Parker, E. T. and Mood, A. N., Some balanced Howell rotations for duplicate bridge sessions, Amer. Math. Mon. 62 (1955), 714-716.

[27] Roselle, K. G., Balancing carry-over effects in round robin tournaments, Biometrika 67 (1980), 127-131.

[28] Room, T. G., A new type of magic square, Math. Gazette 39 (1955), 307.

[29] Scheid, F., A tournament problem, Amer. Math. Mon. 67 (1960), 39-41.

[30] Schellenberg, P. J., On balanced Room squares and complete balanced Howell rotations, Aequationes Math. 9 (1973), 75-90.

[31] Schellenberg, P. J., van Rees, G. H. J. and Vanstone, S. A., The existence of balanced tournament designs, Ars Combin. 3 (1977), 303-318.

[32] Schellenberg, P. J. and Vanstone, S. A., The existence of Howell designs of side $2n$ and order $2n+2$, Proc. 11th S-E Conf. Combin., Graph Thy., Comput., (1980), 879-887.

[33] Stanton, R. G. and Mullin, R. C., Construction of Room squares, Ann. Math. Statist. 39 (1968), 1540-1548.

[34] Stinson, D. R., The existence of Howell designs of odd side, J. Combin. Thy. Ser. A 32 (1982), 53-65.

[35] Wallis, W. D., Spouse-avoiding mixed double tournaments, Ann. NY Acad. Sci., (1979), 549-554.

[36] Wang, S. M. and Wilson, R. M., A few more squares, II, Proc. 9th S-E Conf. Combin., Graph Thy., Comput., (1978), 688.

[37] de Werra, D., Scheduling in Sports, in Studies on Graphs and Sports, ed. P. Hansen, North Holland, (1981), 381-395.

[38] Williams, E. J., Experimental designs balanced for the estimation of residual effects of treatments, Austral. J. Sci. Res. Ser. A 2 (1949), 149-168.

[39] Yalavigi, C. C., A tournament problem, Math. Student 3 (1963), 51-64.

BANDWIDTH IN MULTIGRIDS FOR RANDOM GRAPHS

ZEVI MILLER

Department of Mathematics and Statistics
Miami University
Oxford,Ohio 45056
USA

1. INTRODUCTION AND BACKGROUND

It is the purpose of this article to give a brief introduction to the probabilistic method in graph theory, and to apply this method to a problem motivated by VLSI theory and parallel computation.

Let X be a random variable defined on a probability space Ω. Let $E(X) = \sum_{z \in \Omega} X(z)P(z)$ denote the **expectation** of X and $V(X) = E(X^2) - E(X)^2$ the **variance** of X. Our main tools will be two inequalities involving the expectation and variance.

The first inequality is the following.

Markov's inequality: If $X \geq 0$ and $t > 0$, then $P(X \geq t) \leq \frac{E(X)}{t}$.

This follows directly from the definition of $E(X)$. Frequently one makes use of the following corollary.

Corollary M: $P(X \geq 1) \leq E(X)$.

Now consider a sequence of probability spaces with associated random variables (which we continue to call X). If X is integer valued and nonegative, and if we know that $E(X) \to 0$ in the sequence, then by Corollary M we can conclude that $P(X = 0) \to 1$. In many applications where $X(z)$ is the number of combinatorial configurations of a certain type contained in z, this conclusion says that with probability approaching 1 in the sequence of spaces there exists no configuration of this type in a random element z. Thus Markov's inequality is a tool for proving the almst certain nonexistence of some combinatorial configuration.

By contrast, the second inequality is a tool for proving the almost certain existence of some configuration.

Chebyshev's inequality: If $t > 0$, then $P(|X_E(X)| \geq t) \leq \frac{V(X)}{t^2}$. This inequality can be proved simply by applying Markov's inequality to the random variable $(X - E(X))^2$. Again we often use a corollary.

Corollary C : $P(X = 0) \leq \frac{V(X)}{E(X)^2}$.

Thus if we can show that $\frac{V(X)}{E(X)^2} \to 0$ in our sequence of spaces, then $P(X = 0) \to 0$ in our sequence. Again when X counts the number of configurations of a certain type (so that in particular X is nonnegative integer valued), then we conclude that $P(X \geq 1) \to 1$. this says that with probability approaching 1 there will exist at least one such configuration in a randomly chosen element of our space.

In this paper we will concentrate on the probability space $\boldsymbol{G}(n,p)$ consisting of all labelled graphs having n points with edge probability p. We view each edge as occurring with edge probability p, and these occurrences are assumed to be mutually independent. Thus if $G \in \boldsymbol{G}(n,p)$ has exactly t edges, then the probability of G is $P(G) = p^t(1-p)^{\binom{n}{2}-t}$. The space $\boldsymbol{G}(3,p)$ is illustrated in Figure 1, with probabilities of the graphs indicated. A subset of $\boldsymbol{G}(n,p)$ for some n is called an **event**. The probability of an event A_n of $\boldsymbol{G}(n,p)$ is by definition $P(A_n) = \sum_{G \in A_n} P(G)$. When A_n is an event defined by some structural property of graphs, then we may refer to A_n by naming the property. For example, if $A_n(t)$ is the subset of $\boldsymbol{G}(n,p)$ consisting of all graphs G having exactly t edges, then we may refer to $A_n(t)$ by the phrase " G has t edges". The probability of $A_n(t)$ may then be written $P(G$ has t edges$= \left(\binom{n}{2} \atop t\right)p^t(1-p)^{\binom{n}{2}-t}$ since $A_n(t)$ has $\left(\binom{n}{2} \atop t\right)$ elements, each occurring with probability $p^t(1-p)^{\binom{n}{2}-t}$. Note t probability of the whole space $\boldsymbol{G}(n,p)$ is $\sum P(A_n(t)) = 1$ by the binomial theorem. When A is a property of graphs with arbitrarily many points and A_n is its restriction to $\boldsymbol{G}(n,p)$, then we say that A happens **almost certainly** if $P(A_n) \to 1$ as $n \to \infty$.

We now analyze an event in $\boldsymbol{G}(n,p)$ in more detail. First recall that a point of degree 0 in a graph is called an **isolated point**. A glance at Figure 1 shows that $P(G \in \boldsymbol{G}(3,p)$ has an isolated point$) = (1-p)^3 + 3p(1-p)^2$. Now suppose $p = p(n)$ depends on n, and consider the random variable X on $\boldsymbol{G}(n,p)$ defined by $X(G) =$ the number of isolated points in G. Obviously if $p = 1$ then $P(X(G) = 0) = 1$, while if $p = 0$ then $P(X(G) = 0) = 0$. We may ask for a function $p = p(n)$ which is just barely large enough so that $P(X(G) = 0) \to 1$ as $n \to \infty$. To find this function we will use both corollaries above, one as a tool in a nonexistence proof and the other as a tool in an existence proof.

We begin by defining random variables X_j on $\boldsymbol{G}(n,p)$, $1 \le i \le n$, by $X_j(G) = 1$ or 0 according to whether point i is isolated, or not isolated respectively in G. Hence we have $X = X_1 + X_2 + \cdots + X_n$. First we claim that

$$E(X) = n(1-p)^{n-1}. \tag{1A}$$

To see this, observe that the event that a given point i is isolated requires the $n-1$ nonedges $\{ik : i = k\}$, and an arbitrary configuration of edges on the vertices $\{2, 3, \cdots, n\}$. Hence this event has probability

$$P(X_j = 1) = (1-p)^{n-1} \sum_{t=0}^{\binom{n-1}{2}} \left(\binom{n-1}{2} \atop t \right) p^t (1-p)^{\binom{n}{2}-t} = (1-p)^{n-1}.$$

Since $E(X_i) = P(X_j)$ because X_j is 0-1 valued, the linearity of expectation implies (1A). Now using thee Taylor expansion of the logarithm we find that $E(X) \sim ne^{-np}$ provided $np^2 \to 0$ with n. It follows that if we let $p(n) = c\frac{\log(n)}{n}$ with $c > 1$, then $E(X) \to 0$ as $n \to \infty$, and hence by Corollary M we have $P(X(G) = 0) \to 1$ as $n \to \infty$. We have thus used Corollary M to prove the almost certain nonexistence of isolated points in random elements of $\boldsymbol{G}(n,p)$ when $p = p(n) = c\frac{\log(n)}{n}$ with $c > 1$.

Next we will use Corollary C to prove the almost certain existence of isolated points in random elements of $G(n,p)$ when $p = p(n) = c\frac{\log(n)}{n}$ with $0 < c < 1$. Using linearity of expectation and $X_j^2 = X_j$ for all i, we have

$$E(X^2) = E(X) + \sum_{i \neq k} E(X_j X_k).$$

By symmetry all terms in the sum have the same value, and there are $n(n-1)$ of them. Hence fixing i and k to be 1 and 2 say, the sum becomes $n(n-1)E(X_1X_2) = n(n-1)P(X_1X_2 = 1)$. Now $P(X_1X_2 = 1)$ is just the probability of the event that points 1 and 2 are both isolated. This event requires the $2(n-2)+1$ nonedges $\{1j, 2j, 12 : 3 \leq j \leq n\}$ and an arbitrary configuration of edges on the points $\{3, 4, \cdots, n\}$. Now arguing as before we get $P(X_1X_2 = 1) = (1-p)^P 2(n-2) + 1$. Hence $E(X^2) = E(X) + n(n-1)(1-p)^{2(n-)+1}$. Then using (1A) above we find that

$$\frac{E(X^2)}{E(X)^2} = \frac{1}{E(X)} + \frac{n(n-1)}{n^2} \cdot \frac{1}{1-p}.$$

Now with $p = p(n) = c(\frac{\log(n)}{n})$ and $0 < c < 1$ we have $E(X) \to \infty$ and thus $\frac{E(X^2)}{E(X)^2} \to 1$ as $n \to \infty$. It follows by Corollary C that $P(X(G) = 0) \to 0$ as $n \to \infty$. This proves the almost certain existence of isolated points when $p = p(n) = c\frac{\log(n)}{n}$ and $0 < c < 1$.

In the rest of the paper we will be considering a more complex event and more elaborate asymptotics, but the general method remains the same; Corollaries M and C are used inproving almost certain nonexistence and existence respectively. For further applications of this method and many other topics in random graph theory, the reader is referred to the excellent books [Bo] and [Pa].

2. The problem and its motivation

Let G and H be graphs with $|G| = |H|$, and $f : V(G) \to V(H)$ a one to one map of the point set of G to that of H. Let $|f| = \max[\text{dist}_H(f(x), f(y)) : xy \in E(G)]$ (where dist_H denotes distance in H), and let $B(G,H) = \min_f |f|$.

The investigation of $B(G,H)$ for certain classes of host graphs' H, apart from its intrinsic interest as a graph theory problem, is motivated by issues in VLSI and parallel computation. In VLSI applications the parameter $B(G,H)$ gives a lower bound for the length of wires when an electronic circuit (modelled by G) is embedded on a chip (modelled by H) ([Ro], [RoSn],[Lei]). In parallel computation we may have an algorithm A designed to run on a network of processors G, and instesd we wish to run A on a different network H. Here $B(G,H)$ is a lower bound on the communication delay per unit step when we simulate G by H ([LMS],[MoSu]).

When H is a path the parameter $B(G,H)$ is known as the **bandwidth** $B(G)$ of G. The motivation for studying $B(G)$ arose first in numberical analysis as follows. Given a symmetric $n \times n$ matrix M with O's on the diagonal, one may wish to perform a symmetric permutation of the rows and columns of M with a view to bringing the nonzero entries of the resulting matrix into as narrow band as possible about the diagonal. The reason for doing this is that there are algorithms for certain matrix operations, such as Gaussian eliminatiion and matrix inversion, which work fastest when this band is narrow. Now let $G(M)$ be the graph on n points $\{1, 2, \cdots, n\}$ with i and j joined by an edge if and only if the ij'th entry of M is nonzero. Then $B(G(M))$ is the width of the smallest possible band achievable.

It is well known that the problem of computing $B(G)$ is NP-complete, even when G is a tree of maximum degree three [GGJK]. This has prompted work on the probabilistic analysis of this problem. Turner [Tu] explains the success of some well known heuristics from a probabilistic point of view. Kuang and McDiarmid [KMc] show that for a random graph $G \in \boldsymbol{G}(n, p)$ with fixed edge probability p we have

$$B(G) = n - (2 + 2^{\frac{1}{2}} + o(1))\frac{\log(n)}{\log(\frac{1}{1-p})}$$

with probability approaching 1 as $n \to \infty$. A similar though less precise result appears in [EHW]. Sparse random graphs are considered in [V].

Here we consider the case when H is a multidimensional lattice or grid. Let $[n]^k$ denote the graph with point set $\{x_1, x_2, \cdots, x_k\}$: $0 \leq x_j \leq n, x_j$ integer and an edge between points x and y of $[n]^k$ if and only if $\sum_{i=1}^{k} |x_j - y_j| = 1$. Note that this graph $[n]^k$ has $(n+1)^k$ points and diameter kn. Let $N = (n+1)^k$.

The NP-completeness of determining $B(G, [n]^k)$ when $k = 2$ is shown independently in [Mi], [BCo], and [BSu], and the proofs extend readily to arbitrary dimension k, Bounds for $B(G, [n]^k)$ may be derived from [Ro] and [RoSn], and further work by these and other authors. We wish to investigate the "usual" behavior of $B(G, [n]^k)$.

2. The Main Result

We use the standard notation $O(s(n)), o(s(n))$, and $\Omega(s(n))$ to denote any function $g(n)$ for which the absolute ratio $|\frac{g(n)}{s(n)}|$ is bounded above by a constant, approaches 0, and is bounded below by a constant respectively as n approaches ∞.

We will need the following facts for later use. The first gives standard bounds on binomial coefficients, and the second follows readily from Taylor's theorem.

Lemma 1:

a) $(\frac{n}{k})^k \leq \binom{n}{k} \leq (\frac{ne}{k})^k$ for any $k \leq n$.

b) Suppose $r(k)$ is a function satisfying $r(k) = o(k)$. Then $(1 + \frac{r(k)}{k})^k \sim \exp(O(r(k)))$ as $k \to \infty$.

We now introduce some notation. A point of $[n]^k$ will be called an **extreme point** if each of its coordinates is 0 or n. Note that there are 2^k such points inP $[n]^k$. For each extreme point v let $\boldsymbol{D_i(v)}$ be the set of points in $[n]^k$ at distance i from v and let $\boldsymbol{Nb(v,r)} = \cup_{i=0}^{r} D_j(v)$. We call the sets $D_j(v)$ the **diagonals** at v and $Nb(v, r)$ the **r-neighborhood** about v, and note that by symmetry $|Nb(v, r)| = |Nb(w, r)|$ for any two extreme points v and w. For $i \leq n$ we can identify $D_j(v)$ with the set of nonnegative integer sequences $(x_1, x_2, \cdots, x_k)$ such that $\sum_{j=1}^{k} x_j = i$, so $|D_j(v)| = \binom{i+k-1}{k-1}$ for $i \leq n$. As $|D_i(v)|$ is independent of v, we denote it by $\boldsymbol{D_j}$. Also for $r \leq n$ we then have $|Nb(v, r)| = \sum_{i=0}{}^{r}\binom{i+k-1}{k-1} = \binom{r+k}{k}$.

Let $r \geq 0$ be and consider a sequence of sets $(S_0, S_1, \cdots, S_r)$ where S_i is a subset of $G \in \boldsymbol{G}(n,p)$ (with $N = (n+1)^k$) of size D_j, and $S_i \cap S_j = \emptyset$ for $i \neq j$. The set $\cup_{i=0}^{r} S_j$ will be called an **r-corner** of G (or just corner when r is understood). Thus an r-corner in G has the same size as an r- neighborhood about an extreme point in $[n]^k$, and we will refer to the sets S_j as the **diagonals** of the corner. Now let $v(1), v(2), \cdots, v(2^k)$ be a fixed ordering of the extreme points of $[n]^k$. A sequence $(S^{(1)}, S^{(2)}, \cdots, S^{(2^K)})$ of **r-corners** of G will then be called an r- system if the hypergraphs $\cup_{j=0}^{2^k} S_j$ and $\cup_{j=0}^{2^k} Nb(v(j), r)$ (with edge sets $\{S^{(j)} : 1 \leq j \leq 2^k\}$ and $\{Nb(v(j), r) : 1 \leq j \leq 2^k\}$ respectively) are isomorphic by an isomorphism under which the induced map on edges sends $S^{(j)}$ to $Nb(v(j), r)$ for each j. As motivation for these definitions note that if $f : G \to [n]^k$ is an embedding, then the inverse image of the collection $[Nb(v,r) : v \text{ extreme}]$ must be an r-system of G while the inverse image of each $Nb(v,r)$ must be an r-corner of that r-system.

We will assume now that the ordering $v(i)$ of the extreme points is such that for any odd i the pair $[v(i), v(i+1)]$ are antipodal (i.e. at distance kn). Let S and T be two r-corners of G (with diagonal sets $\{S_j\}$ and $\{T_j\}$ respectively) such that each point $v \in S_j$ is nonadjacent in G to all points in $\cup_{j=0}^{r-i} T_j$. (It follows symmetrically that any point $v \in T_j$ is nonadjacent in G to all points in $\cup_{j=0}^{r-i} S_j$). We then call S and T **opposite corner**. An r-system will then be termed a **goodr-system** if each pair $\{S^{(i)}, S^{(i+1)}\}$ with i odd are opposite corners. If S is a good r-system of G, then the required set of nonadjacencis among the 2^{k-1} opposite corner pairs of S will be referred to as the **nonadjacencies implied** by S.

The significance of a good r-system is seen in the following lemma.

Lemma 2 : $B(G) \leq \text{diam}([n]^k) - r - 1 \Leftrightarrow G$ has a good r-system.

Proof : Observe that points $s, t \in [n]^k$ satisfy $\text{dist}(s,t) \geq kn - r$ iff there are antipodal extreme points $v(i)$ and $v(i+1)$ such that (without loss of generality) $S \in D_j(v)$ and $t \in \cup_{t=0}^{r-j} D_t(v(i+1))$ for some $j \leq r$.

Now let $f : G \to [n]^k$ be an embedding. It follows that $|f| \leq kn - r - 1$ if and only if for each extreme point $v(i) \in [n]^k$ (say with i odd) the two sets $f - 1(Nb(v(i), r))$ and $f^{-1}(Nb(v(i+1), r))$ are opposite corners in G. Thus $|f| \leq kn - r - 1$ if and only if the sequence $(f^{-1}(Nb(v(1), r)), f^{-1}(Nb(v(2), r)), \cdots, f^{-1}(Nb(v(2^k), r)))$ is a good r-system in G. QED

We may now proceed to our theorem.

Theorem : Let $N = (n+1)^k$. Assume $\Omega(k) = n \leq e^{e^{O(k)}}$. Then there are constants C_1 and C_2 such that for p fixed and $G \in \boldsymbol{G}(N,p)$ we have $P[\text{diam}([n]^k) - C_1 \log(\log N) \leq B(G) \leq \text{diam}([n]^k) - C_2 \frac{\log(\log N)}{\log k}] \to 1$ as $k \to \infty$.

Toward proving this theorem we begin with a lemma.

Lemma 3 : The number of nonadjacencies implied by an r-system for $r \leq n$ is $2^{k-1}\binom{2k+r}{r}$.

Proof : Consider an opposite corner pair S and T (with diagonal sets $\{S_i\}$ and $\{T_i\}$ respectively) belonging to an r-system. For a given $i \leq r$ we have

$$|\cup_{m=0}^{r-i} T_m| = \sum_{m=0}^{r-i} \binom{m+k-1}{k-1} = \binom{r-i+k}{k}.$$

Thus the number of nonadjacencies between S_j and $\cup_{m=0}^{r-i} T_m$ is $\binom{i+k-1}{k-1}\binom{r-i+k}{k}$. Hence the number of implied nonadjacencies among S and T is

$$\sum_{i=0}^{r}\binom{i+k-1}{k-1}\binom{r-i+k}{k}=\sum_{i=0}^{r}\binom{i+k-1}{i}\binom{r-i+k}{r-1}=\binom{2k+r}{r}.$$

Since there are 2^{k-1} pairs of opposite corners in an r-system, the lemma follows. QED

As further notation, for $G \in \boldsymbol{G}(N,p)$ let $N_r(G)$ be the number of r-system in G. Also if $n = rs$ then let $\binom{n}{r\times s}$ denote the multinomial coefficient $\binom{n}{r,r,\cdots,r}$ where the r appears s times in the bottom.

Proof of theorem: First we show that $P[B(G) > \operatorname{diam}([n]^k) - C\log(\log N)] \to 1$ as $k \to \infty$.

Let $N_r(G)$ be the number of good r-systems in $G \in \boldsymbol{G}(n,p)$. We will begin by showing that when $r = C\log(\log N)$ for a suitable constant C, then the expectation of N_r satisfies $E(N_r) \to 0$ as $k \to \infty$.

First observe that with $r = C\log(\log N)$ and k sufficiently large we have, letting $q = 1-p$,

$$E(N_r)=\binom{n}{2^k\binom{r+k}{k}}\binom{2^k\binom{r+k}{k}}{\binom{r+k}{k}\times 2^k}$$
$$\binom{\binom{r+k}{k}}{\binom{k-1}{k-1},\binom{1+k-1}{k-1},\cdots,\binom{r+k-1}{k-1}}^{2^k}(2^k)!\,q^{\binom{r+2k}{r}2^{k-1}}.$$

To see this, note that since $r = \log(k) + \log\log(n) = o(k)$ and $n = \Omega(k)$ it follows that for k sufficiently large the r-corners making up an r-system are pairwise disjoint. Hence the number of points in an r-system for large k is $2^k\binom{r+k}{k}$. Thus the first factor on the right is the number of ways of choosing the ground set of $2^k\binom{r+k}{k}$ points for our r-system. The second factor is the number of ways to partition these points into a collection of 2^k corners. The third factor is the number of ways to partition each of the corners into the diagonals $S_0, S_1, S_2, \cdots, S_r$, these partitions being independent. The fourth factor is the number of different orderings of the 2^k corners. Thus the product of the first four factors is the number of r-systems. The exponent of q is the number of disallowed edges in a good r- system by Lemma 3, so the fifth factor is the probability that an r- system is a good r-system. Thus the full product is the expectation of N_r.

We now estimate $E(N_r)$. First we will bound the multinomial coefficients by powers of binomial coefficients. The one with $2^k\binom{r+k}{k}$ as the top is clearly at most

$$\binom{2^k\binom{r+k}{k}}{\binom{r+k}{k}}^{2^k}.$$

The one with $\binom{r+k}{k}$ as the top may be written as the product

$$\prod_{j=0}^{r}\left(\frac{\binom{j+k}{k}}{\binom{j+k-1}{k-1}}\right)$$

using the fact that for any integer $j \geq 0$ we have

$$\sum_{i=0}^{j}\binom{i+k-1}{k-1} = \binom{j+k}{k}.$$

Now some analysis of binomial coefficients shows that the biggest among the $r+1$ factors in this product is

$$\left(\frac{\binom{r+k}{k}}{\binom{r+k-1}{k-1}}\right)$$

Hence the multinomial coefficient is bounded above by

$$\left(\frac{\binom{r+k}{k}}{\binom{r+k-1}{k-1}}\right)^{r+1}$$

Now substituting for the multinomial coefficients by the above bounds, using Lemma 1 to upper bound all binomial coefficients, and applying the Stirling approximation to $(2^k)!$, we get $E(N_r) \leq M^{2^k}$, where

$$M = \left(\frac{Ne}{2^k\binom{r+k}{k}}\right)^{\binom{r+k}{k}} (2^k e)^{\binom{r+k}{k}} \left(\frac{\binom{r+k}{k}e}{\binom{r+k-1}{k-1}}\right)^{r+1\binom{r+k-1}{k-1}} \cdot F \cdot q^{\binom{2k+r}{r}}$$

and $F = \frac{2^k}{e}\pi 2^{-(k+1)}$, and a factor of $\frac{1}{2}$ in the exponent of q has been dropped as it does not affect the asymptotics. The statement $E(N_r) \to 0$ as $k \to \infty$ would therefore follow if we knew that $M \leq 1 - \delta$ for some arbitrarily small but fixed $\delta > 0$.

We will show that $M \leq 1 - \delta$ holds by starting with $M \leq 1 - \delta$ and working backwards to our assumption on r. Taking logarithms we get

$$\binom{r+k}{k}[k\log(n+1) + 2 - \log(\binom{r+k}{k})] + (r+1)\binom{r+k-1}{k-1}[\log(\frac{r+k}{k}) + 1]$$
$$+\log(F) + \binom{2k+r}{r}\log(q) < -\epsilon, \tag{3A}$$

where $\epsilon = |\log(1-\delta)|$ (so ϵ is fixed and small when δ is small). We are reduced to showing that (3A) is true for sufficiently large k.

Clearly $\log(F) = O(k)$ and hence is dominated by $\binom{r+k}{k}k\log(n+1)$ for large k. Thus $\log(F)$ may be eliminated after inserting an appropriate constant in front of the latter. Also note that if the inequality resulting from (3A) by eliminating all terms from the first brackets except $k\log(n+1)$ were true, then (3A) would be true. We are then reduced to proving the inequality below, call it (3B), obtained by eliminating these terms and $\log(F)$.

$$\binom{r+k}{k}[k\log(n+1)]+(r+1)\binom{r+k}{k-1}[\log\frac{r+k}{k}+1]+\binom{2k+r}{r}\log(q)<-\epsilon. \quad (3B)$$

Now divide both sides of (3B) by $\binom{r+k}{k}$. Simplification gives

$$\frac{\binom{r+k-1}{k-1}}{\binom{r+k}{k}}=\frac{k}{r+k}.$$

Also we have (employing the "falling factorial" notation $a_{(c)}=a(a-1)(a-2)\cdots(a-c+1)$)

$$\frac{\binom{r+2k}{r}}{\binom{r+k}{k}}=\frac{(r+2k)_{(k)}}{2k_{(k)}}=\prod_{i=0}^{k}(1+\frac{r}{2k-1})\geq(1+\frac{r}{2k})^{k}=O(1)\exp(\frac{r}{2}) \text{ as } k\to\infty,$$

using $r=o(k)$ and Lemma 1. We are therefore reduced to showing thet

$$Ck\log(n+1)+(\frac{r+1}{r+k}k[\log(\frac{r+k}{k}+1]+C'\exp(\frac{r}{2})\log(q)<\frac{-\epsilon}{\binom{2r+k}{k}} \quad (3C)$$

for large k, where C and C' are constants.

Now observe that since $r=o(k)$ we have $r<Ckn$ for k sufficiently large and any constant C. Hence $\log(\frac{r+k}{k})\ll\log(n)$ (with increasing k) so that the middle term of (3C) may be eliminated by adjusting the constant C in the first term. We are thus led to showing that

$$Ck\log(n+1)+C'\exp(\frac{r}{2})\log(q)<-\epsilon,$$

where ϵ is and arbitrarily small positive constant. The last inequality is equivalent to

$$C'\exp(\frac{r}{2})>\frac{\epsilon+Ck\log(n+1)}{\log(\frac{1}{q})}.$$

Finally we see that this would be implied if $r>C_1\log(k\log(n+1))=C_1\log(\log N)$ where C_1 is an appropriate constant.

We have thus seen that $E(N_r)\to 0$ as $k\to\infty$ for $r=C\log(\log N)$ with a suitable constant C. Now using Lemma 2 and Markov's inequality we have

$$P[B(G)\leq \operatorname{diam}([n]^k)-r-1]=P[N_r\geq 1]\leq E(N_r)\to 0$$

as $k\to\infty$ for $r=C\log(\log N)$. Hence $P[B(G)>\operatorname{diam}([n]^k)-C\log(\log N)]\to 1$ as $k\to\infty$, proving half of our theorem.

To prove that

$$P[B(G) \leq \operatorname{diam}([n]^k) - C_2 \frac{\log(\log N)}{\log k}] \rightarrow 1$$

we must prove the almost certain existence of an appropriate r-system by Lemma 2, and this will be done by applying the Corollary C of Chebyshev's inequality.

Let R be the set of r-systems. For each $A \in R$ let

$$Z_A(G) = \begin{cases} 1, & \text{if } A \text{ is a good } r\text{-system} \\ 0, & \text{otherwise.} \end{cases}$$

Note that $N_r(G) = \sum_{A \in R} Z_A(G)$.

We begin by estimating the variance of N_r. Let F be a fixed member of R. Then

$$\begin{aligned} \operatorname{Var}(N_r) &= E(N_r^2) - E(N_r)^2 \\ &= \sum_{A \in R} \sum_{B \in R} (E(Z_A Z_B) - E(Z_A)E(Z_B)) \\ &= |R| \sum_{A \in R} (E(Z_F Z_A) - E(Z_F)E(Z_A)). \end{aligned}$$

To estimate $E(Z_F Z_A)$, note first that since Z_F and Z_A are 0-1 valued random variables we have $E(Z_F Z_A) = P(Z_F Z_A = 1) = P(Z_A = 1 | Z_F = 1) \cdot P(Z_F = 1)$. By Lemma 3 we have $P(Z_F = 1) = q^{\binom{2k+r}{r} 2^{k-1}}$. Now given that F is a good r-system the probability that A is a good r-system is just the probability that A has the required nonadjacencies not already implied by F. This probability is $q^{\binom{2k+r}{r} 2^{k-1} - c(A)}$, where $c(A)$ is the number of nonadjacencies implied jointly by A and F. Thus

$$E(Z_F Z_A) = q^{\binom{2k+r}{r} 2^k - c(A)}.$$

We can now develop the above expression a bit further. Let $h = 2^k \binom{r+k}{k}$, and let ν_t be the set of elements of R having t points in common with F. Also let $c(t)$ be the maximum number of nonadjacencies implied jointly by F and any member of ν_t. Then continuing from above we have

$$\begin{aligned} \operatorname{Var}(N_r) &= |R| q^{\binom{2k+r}{k} 2^k} \sum_{A \in R} (q^{-c(A)} - 1) \\ &\leq |R| q^{\binom{2k+r}{r} 2^k} \sum_{t=2}^{h} \| \nu_t | q^{-c(t)} \\ &= |R|^2 q^{\binom{2k+r}{r} 2^k} \sum_{t=2}^{h} \frac{|\nu_t|}{|R|} q^{-c(t)} \\ &= E(N_r)^2 \sum_{t=2}^{h} \frac{|\nu_t|}{|R|} q^{-c(t)}. \end{aligned}$$

We therefore have

$$\frac{\mathrm{Var}(N_r)}{E(N_r)^2} \leq \sum_{t=2}^{h} \frac{|\nu_t|}{|R|} q^{-c(t)}. \tag{3D}$$

It remains to estimate the right side of the last expression. First note that we may write $\nu_t = N_1 N_2$, where

N_1 =number of ways to choose h points from $(1, 2, \cdots, N)$ of which t are in the ground set of F, and

N_2 =number of ways to arrange any gound set counted in N_1 into a member of R.

We have $N_1 = \binom{h}{t}\binom{N-h}{h-t}$, while $N_2 = \frac{|R|}{\binom{N}{h}}$. Therefore from (3D) we get

$$\frac{\mathrm{Var}(N_r)}{E(N_r)^2} \leq \sum_{t=2}^{h} \frac{\binom{h}{t}\binom{N-h}{h-t}}{\binom{n}{h}} q^{-c(t)}. \tag{3E}$$

We recognize the coefficients of the $q^{-c(t)}$ in this sum as the terms of the hypergeometric distribution. In the course of approximating the hypergeometric by the Poisson distribution one can derive the following inequality [Bo pp.7-8].

$$\frac{\binom{h}{t}\binom{N-h}{h-t}}{\binom{N}{h}} \leq \frac{(\frac{h^2}{N})^t \exp(-\frac{h^2}{N})}{t!}(1 - \frac{t}{N})^{t-h} \exp(\frac{ht}{N})$$

(This is obtained by first approximating the hypergeometric by the binomial distribution $b(t; h, p)$ with success probability $p = \frac{h}{N}$, and then the binomial by the Poisson distribution with mean $\lambda = ph = \frac{h^2}{N}$.)

Next we need an estimate for $c(t)$. Consider then a set $Q \in \nu_t$ at which the maximum defined by $c(t)$ is realized. Thus $c(t)$ is the number of nonadjacent pairs $xy, x, y, \in Q \cap F$, implied jointly by Q and F. Now $c(t) = \frac{1}{2}\sum_{x \in Q \cap F} d(x)$, where $d(x)$ is the number of nonadjacencies implied jointly by Q and F in which x is a member. Since $Q \in R$ we have $d(x) \leq \binom{r+k}{k}$, with equality occurring precisely when there is a corner pair S, T in Q for which $x = S_0$ and the points counted in $d(x)$ comprise all of T. Thus $c(t) \leq \frac{1}{2}t\binom{r+k}{k}$.

Now using Corollary C and (3E) we therefore have

$$\begin{aligned} &P(B > \mathrm{diam}([n]^k) - r - 1) = P(N_r = 0) \leq \frac{\mathrm{Var}(N_r)}{E(N_r)^2} \\ &\leq \exp(-\frac{h^2}{N}) \sum_{t=2}^{h} \frac{(\frac{h^2}{N})^t}{t!}(1 - \frac{t}{N})^{t-h} \exp(\frac{ht}{N}) q^{-t\binom{r+k}{k}} \\ &\leq \exp(-\frac{h^2}{N})(1 - \frac{h}{N})^{-h}[\exp\{\frac{h^2}{N}\exp(\frac{h}{N}) q^{-\binom{r+k}{k}}\} - 1 - \frac{h^2}{N}\exp(\frac{h}{N}) q^{-\binom{r+k}{k}}]. \end{aligned} \tag{3F}$$

Recalling that $h = 2^k\binom{r+k}{k} \leq ((1 + \frac{r}{k})2e)^k$ (by Lemma 1), $r \leq C\frac{\log\log(N)}{\log(k)}$, $N = (n+1)^k$, and the bound on n from the hypothesis, it follows readily that $h^2 = o(N)$

as $k \to \infty$. Hence we get $(1-\frac{h}{N})^{-h} \to 1, \exp(\frac{h}{N}) \to 1$, and $\exp(-\frac{h^2}{N} \to 1$ as $k \to \infty$. Thus a condition sufficient to make expression (3F) approach 0 as $k \to \infty$ would be

$$q^{-\binom{r+k}{k}} \leq \left(\frac{N}{h^2}\right)^{1-\epsilon} \tag{3G}$$

for some arbitrarily small but fixed $\epsilon > 0$.

The proof of this inequality under the assumption on r is straightforward, but we include it here for completeness. Using Lemma 1 we have $\binom{r+k}{k} \leq ((1+\frac{k}{r})e)^r$, and taking logarithms we get

$$((1+\frac{k}{r})e)^r \log(\frac{1}{q}) \leq C \log(N)$$

for some constant C, where we have used the fact (seen above) that h^2 is small compared to N. Taking logarithms again and noting that $1+\frac{k}{r} \leq k$ for large enough k (and hence r), it follows that if $r \leq C\frac{\log\log(N)}{\log(k)}$ (where the constant C has been adjusted) then (3G) holds. Since this is our assumption on r and these steps are reversible, we now have (3G).

We have thus shown that $P(B > \text{diam}([n]^k) - r - 1) \to 0$ as $k \to \infty$ for $r \leq C\frac{\log\log(N)}{\log(k)}$. It follows that almost every graph satisfies $B(G) \leq \text{diam}([n]^k) - C\frac{\log\log(N)}{\log(k)}$ and the theorem is proved. QED

Remark: The main theorem of this paper has just recently been improved in [McMil].

References

[Bo] B.Bollobas,*Random Graphs*, Academic Press (1985).

[BCo] S.Bhatt, S.Cosmadakis, "The complexity of minimizing wire lengths in VLSI layouts", manuscript (1983).

[BSu] P.Bertolazzi, I.H.Sudborough, "The grid embedding problem is NP-complete even for edge length 2", Technical Report (1983), EE/CS Dept., Northwestern University.

[EHW] P.Erdos, P.Hell, P.Winkler, "Bandwidth versus bandsize", *Graph Theory at Sandbjerg Castle-Proceedings of the Conference in Memory of Gabriel Andrew Dirac* June 1985.

[GJK] M.Garey, R.Graham, D.Johnson, D.Knuth, "Complexity results for bandwidth minimization", *SIAM J. of Applied Mathematics*, 34 (1978), 477-495.

[KMc] Y.Kuang, C.McDiarmid, "On the bandwidth of a random graph", *Ars Combinatoria* 20A(1985), 29-36.

[Lei] F.T.Leighton "A framework for solving VLSI graph layout problems",*J. of Computer and System Sciences*, **28**:2(1984), 300-343.

[LMS] F.T.Leighton, F.Makedon, and I.H.Sudborough, "Simulating pyramids with hypercubes", manuscript in preparation.

[Mc] C.McDiarmid, "Achromatic numbers of random graphs",*Math. proc. Camb. Phil. Soc.*, **92**(1982), 21-28.

[McMil] C.McDiarmid, Z.Miller, "Lattice bandwidth in random graphs", submitted for publication (1988).

[Mil] Z.Miller, J.Orlin, "NP-completeness for minimizing maximum edge length in grid embeddings", *J. of Algorithms* , 6(1985), 10-16.

[MoSu] B.Monien, I.H.Sudborough, "Simulating binary trees on hypercubes", *VLSI Algorithms and Architectures*, Lecture Notes in Computer Science, Springer Verlag (1988), 170-180.

[Pa] E.M.Palmer, *Graphical Evolution: An Introduction to the Theory of Random Graphs*, John Wiley & Sons (1985).

[Ro] A.Rosenberg, "On embedding graphs in grids", IBM Research Report RC 7559 (#2668) (1979).

[RoSn] A.Rosenberg, L.Snyder, "Bounds on the costs of data encodings", *Math. Systems Theory*, **12**(1978), 9-39.

[Tu] J.Turner, " On the probable performance of heuristics for bandwidth minimization", *SIAM J. of Computing*, **15**(1986). 561-580.

[V] W.F.de la Vega, "On the bandwidth of random graphs",*Annals of Discrete Mathematics*, **17**(1983), 633-638.

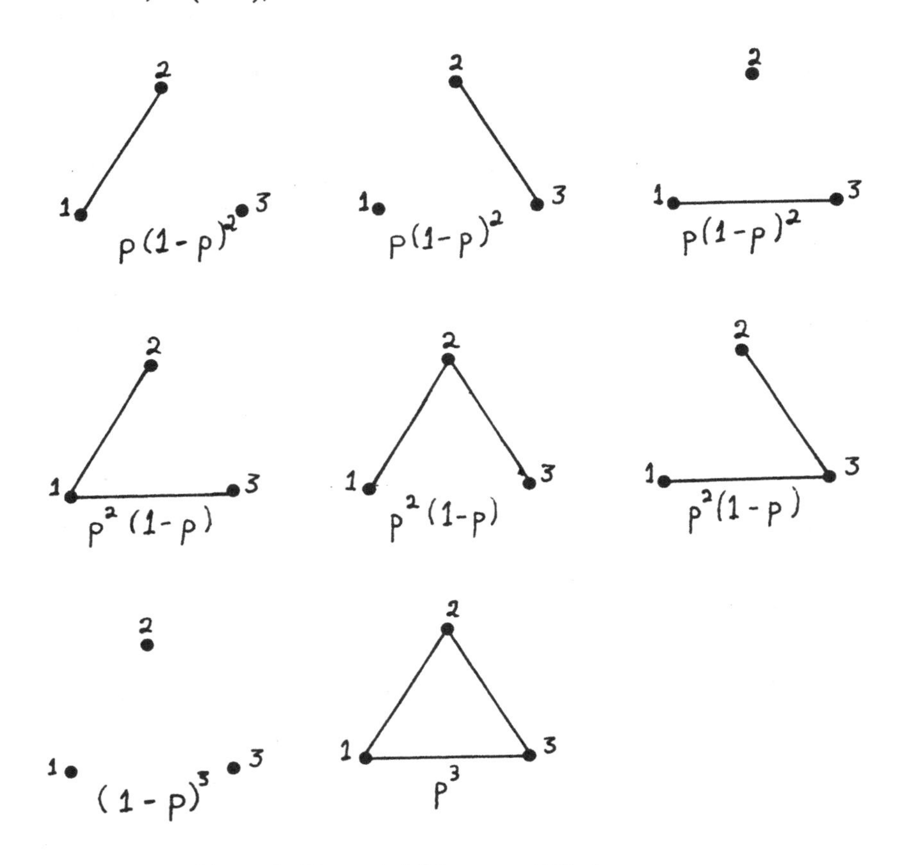

Figure 1: The probability space **G**(3,p)

A FLAVOR OF MATROIDS, GRAPHS AND OPTIMAL JOB ASSIGNMENT PROBLEMS IN OPERATIONS RESEARCH

J.M.S. Simões-Pereira

Departamento de Matemática
Universidade de Coimbra
P-3000-COIMBRA-PORTUGAL

1. INTRODUCTION

Our purpose is to give the reader a feeling for some uses of matroid theory in operations research without assuming previous knowledge of matroids, operations research or any mathematical background other than the basic definitions of graph theory. Graphs and matroids are two of the most important objects in combinatorics. Combinatorial optimization is fundamental in operations research.

We want to appeal to readers who have already been exposed to either matroids or operations research but not both. The topics we chose are suitable for an unsophisticated although rigorous presentation. This was the main reason for our choice. Another reason is that they lead to a vast new area widely open to research.

The proofs we give are all elementary and constitute a coherent set of proofs for results which have been originally studied within distinct settings and using different although equivalent definitions of a matroid as a starting point.

We start with some basic definitions and examples of matroids in Section 2, where we also introduce the problem of finding the maximum or minimum spanning tree of a connected graph with weighted edges. In Sections 3 and 4, we give a general setting for the preceding problem: first we define optimality among subsets of a linearly ordered set and present the Kruskal's algorithm to determine optimal subsets, if they exist; then we give a proof that the collection of independent sets of a matroid always has an optimal set. In Section 5, we apply this result to an assignment problem of operations research. In Section 6, we introduce a class of functions which are known to generate matroids and, in Section 7, using these functions, we associate a certain class of matroids to hypergraphs. In Section 8, we deal with the particular case where the hypergraphs are simple graphs and, in Section 9, hypergraphs and their associated matroids are used to study a generalization of the previously discussed assignment problem. Finally, in Section 10, we refer to some research problems.

2. MATROIDS: A DEFINITION AND EXAMPLES. BASES AND AN OPTIMIZATION PROBLEM.

Among the many equivalent definitions of a finite matroid, we choose the one based on the concept of an independent set. As usually, we will denote the empty set by $\emptyset$ and the cardinality of a set S by $|S|$.

Definition 2.1.

Let E be a finite set, $\mathcal{J} \in P(E)$ a collection of subsets of E which satisfy the following axioms:

J1) $\emptyset \in \mathcal{J}$;

J2) $A \in \mathcal{J} \wedge B \subset A \Rightarrow B \in \mathcal{J}$;

J3) $A, B \in \mathcal{J} \wedge |A| = |B| + 1 \Rightarrow \exists x \in A-B : B \cup (x) \in \mathcal{J}$.

Then the pair $(E, \mathcal{J})$ is a matroid and the sets in $\mathcal{J}$ are the independent sets of the matroid.

An independent set which is maximal among the independent sets with respect to set inclusion is called a <u>base</u>. A set which is not independent is called <u>dependent</u>. A dependent set which is minimal among the dependent sets with respect to set inclusion is called a <u>circuit</u>.

As a consequence of axiom J3, it is easy to show that all bases of a matroid have the same cardinality. We leave the proof to the reader. On the contrary, in general, not all circuits of a matroid have the same cardinality.

The most trivial example of a matroid is the so-called <u>uniform matroid</u>: Given a finite set E, with $|E| = n$, and an integer k, with $0 \leq k \leq n$, the subsets X of E with cardinality $|x| \leq k$ are the independent sets of the k-uniform matroid. The bases of this matroid are the subsets of E with cardinality k; the circuits are the subsets of E with cardinality $k + 1$.

Another example: Given a simple graph $G = (V,E)$, i.e. a graph with no loops nor multiple edges, with V and E as vertex and edge sets, respectively, the subsets of E which form subgraphs of G with no closed paths are the independent sets of a matroid on E. The bases of this matroid are the spanning forests of G (spanning trees, if G is connected). The circuits are the simple closed paths or cycles of G. This matroid is called the polygon matroid of G. In the graph of figure 2.1, {a,b,c}, {a,b,d} and {a,c,d} are the bases and {b,c,d} is the only circuit; besides the bases and the empty set, any set with one or two edges is independent.

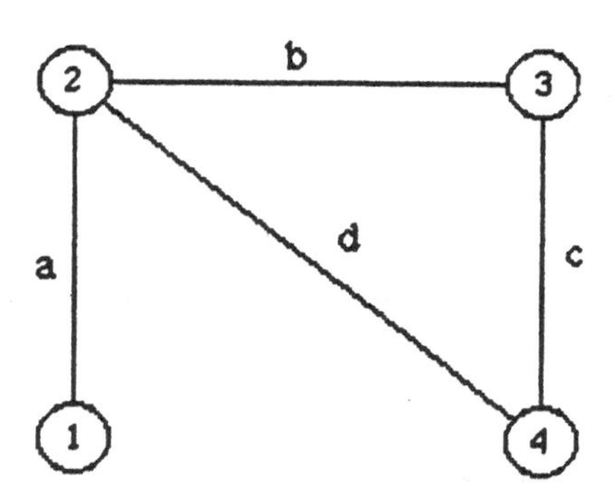

Fig. 2.1

Quite often, the edges of a graph are assigned weights, this means, numerical values which may represent lengths, costs, times or capacities, for instance. We can think of a road map where the edges are road sections connecting towns or intersections. We can also think of electrical networks, where the weights may stand for electrical variables such as voltage differences, or intensity of current. There is a problem which occurs very frequently: it consists in choosing, among all spanning trees of a weighted (connected) graph, the one (or the ones) with minimum or maximum total weight. For example, in the graph of figure 2.2, with the edge weights as given in the figure, the set of edges {g,a,d,e} forms a minimum spanning tree with total weight $s = 12$. Note that, although the weight of c is smaller than the weight of e, the edge c can not appear together with edges a and d for they would form a closed path. Similarly, {f,b,e,d} forms a maximum spanning tree with total weight $s = 27$; here again, although the weight of c is bigger then the weight of d, edge c can not appear together with edges b and e for they would form a closed path.

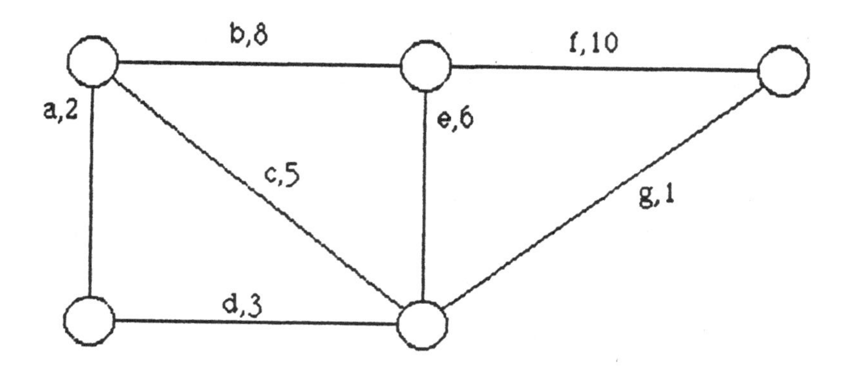

Fig. 2.2

3. OPTIMALITY AMONG SUBSETS OF A LINEARLY ORDERED SET AND THE GREEDY ALGORITHM

To put now these problems into a more general setting, note first that, in our examples, we have a linearly ordered set or chain $(E, \succ)$ where the order relation $\succ$ is given by the order relations $\leq$ or $\geq$ among the numerical weights of the edges. Note also that we have a collection $\mathfrak{F}$ of subsets of E which are, in the same examples, the edge sets of the spanning trees and we want to choose, among the sets of $\mathfrak{F}$, one (or several) which we call optimal. We formalize a general definition of this kind of optimality as follows:

Definition 3.1:

Let $(E, \succ)$ be a finite chain and $\mathfrak{F} \subset P(E)$. A set $A^* \in \mathfrak{F}$ is called <u>optimal</u> if, for any other $B \in \mathfrak{F}$, there is an injective function $f : B \rightarrow A^*$ such that $f(b) \succ b$ for each $b \in B$.

This definition requires only the existence of a linear order relation on E, with no presumption that there are numerical values where a sum is defined. However, the reader is urged to verify that the spanning trees of minimum (or maximum) total weight are optimal also according to this definition. Note also that, $a \succ b$ means "a follows b".

In general, given an arbitrary collection $\mathfrak{F}$ of subsets of a chain $(E, >)$, there does not always exist an optimal element of $\mathfrak{F}$. It exists, for instance, when $\mathfrak{F}$ is the collection of independent sets of a matroid on E.

Before proving this result let us set up our notations and introduce some more concepts.

As already said, $(E, >)$ denotes a finite chain and $\mathfrak{F}$ is a collection of subsets of E. If $A = \{a_1, ..., a_i\} \in \mathfrak{F}$, then, with no loss of generality, we suppose $a_1 > ... > a_i$; similarly, for $B = \{b_1, ..., b_j\} \in \mathfrak{F}$, we suppose $b_1 > ... > b_j$. We say that $A > B$ when we have either $a_k > b_k$ in the first pair a_k, b_k where $a_k \neq b_k$ or $i > j$ and $a_k = b_k$ in all pairs a_k, b_k for $1 \leq k \leq \min(i,j) = j$; the linear order relation that we define on $\mathfrak{F}$ in this way is called lexicographic. If one of the sets in $\mathfrak{F}$ is optimal in the sense of Definition 3.1, then it is the last one in the lexicographic order. Trivially, $\mathfrak{F}$ has always a last set in the lexicographic order. However, $\mathfrak{F}$ has not always an optimal set. For an example, let E be the edge set of the graph in figure 2.2, $>$ the relation $\geq$ among the weights of E and $\mathfrak{F}$ the collection of the sets $\{e, d, g\}$, $\{b, e, c\}$, $\{e, d, a\}$ and $\{f, e, d\}$ whose weights are $\{6,3,1\}$, $\{8,6,5\}$, $\{6,3,2\}$ and $\{10,6,3\}$, respectively; the set $\{f, e, d\}$ is the last one in the lexicographic order as we defined it, but it is not optimal, as we can see by considering the set $\{b, e, c\}$.

When there is an optimal set in $\mathfrak{F}$, we can find it using the following algorithm which was first used by Kruskal [6] to find the spanning tree of a graph with minimum weight. This algorithm was later generalized and became known as the greedy algorithm (See Welsh [15] and Klee [5]):

Algorithm 3.2

- First choose among the elements of E which belong to at least one set in $\mathfrak{F}$ the last one with respect to $>$. Let it be x_1. For the remaining elements of the optimal set, suppose we have already $x_1, ..., x_h$. We choose now the last element with respect to $>$, say x_{h+1}, such that $\{x_1, ..., x_h, x_{h+1}\} \subset A \in \mathfrak{F}$. When no additional element may be chosen, we have the optimal set of $\mathfrak{F}$ we were looking for.

4. OPTIMALITY AMONG THE INDEPENDENT SETS OF A MATROID

The following is the result we previously announced:

Theorem 4.1: Let $(E, >)$ be a finite chain and let $\mathfrak{F}$ be the collection of independent sets of a matroid on E. Then $\mathfrak{F}$ contains an optimal set.

Proof. Let $A = \{a_1, ..., a_n\}$ be the base of the matroid $(E, \mathfrak{F})$ which is, among the bases, the last one in the lexicographic order: since all bases have the same cardinality, this means that, if $B = \{b_1, ..., b_n\}$ is another base, then, for the smallest i such that $a_i \neq b_i$ we have $a_i > b_i$.

We claim that A is optimal. In fact, let $C = \{c_1, ..., c_m\} \in \mathfrak{F}$ and suppose, with no loss of generality, $c_1 > ... > c_m$. By the definitions we have $m \leq n$. If A is not optimal, then there exists C such that $c_r > a_r$ for some r: let r be the smallest subscript which makes $c_r > a_r$. Suppose first that $r > 1$. Since $\{a_1, ..., a_{r-1}\}$ is independent (because it is a subset of A), and $\{c_1, ..., c_r\}$ is independent (because it is a subset of C), the axiom J3 guarantees the existence of $c_s \in \{c_1, ..., c_r\}$ such that $\{a_1, ..., a_{r-1}, c_s\}$ is independent. Now, $c_s > c_r > a_r$. This means that there is an independent set which comes after A in the lexicographic order and, therefore, there is also a base which contains it and hence comes also after A, a contradiction. The same reasoning holds when $r = 1$, the only non-essential difference being that $\{a_1, ..., a_{r-1}\} = \emptyset$ and we may choose $c_s = c_r$.

The theorem is thus proved.

5. APPLICATIONS TO AN ASSIGNMENT PROBLEM.

Theorem 4.1 can be applied in a certain type of job assignment problem (see Gale [2]): in this problem, the given set of jobs is linearly ordered and the order relation $a > b$ between jobs a and b means that "a is more important than b" or "a is more profitable than b", or something similar.

In symbols, let us suppose that we are given a set of jobs $X = \{x_1, ..., x_n\}$ with a linear ordering which corresponds to the importance of the jobs, and a set of workers $Y = \{y_1, ..., y_m\}$; each worker is qualified to execute one or more of the jobs or, in symbols, besides X and Y, we have a qualification function $\Phi : X \to P(Y)$ where $y \in \Phi(x)$ means that the worker y is qualified to do the job x.

In general, we will not be able to have all jobs done; the problem consists in choosing a subset of X which is assignable, that means, which can be assigned to the workers within the limits set by their qualifications and which is important among all assignable subsets of X. By now it should be clear to the reader that we call a subset of jobs A more important than another subset of jobs B when B precedes A in the lexicographic order arising from the linear ordering "more importan than" among the jobs in X.

Is there always an assignable set which we would call the optimal one, in the sense of Definition 3.1.? Obviously, an optimal assignable set, if it exists, is the most important one, that means, the last one in the lexicographic order. Since the last one in the lexicographic order is not necessarily optimal in the sense of Definition 3.1, can we be sure that an optimal one actually exists?

Before going any further, let us assume that we have four jobs such that the first one is more important than the second, the second more important than the third and the third more important than the fourth; and we have two workers, T and U, and a qualification function such that only two subsets are assignable, one formed by the first and fourth jobs, the other formed by the second and third jobs. In this case none of these two sets is optimal, although one is more important than the other one.

This situation however cannot occur: if, in one assignment, we give the first job to T and the fourth to U, then, in the other, we either give the second to T and the third to U (which means that the first and third also form an assignable set) or give the second to U and the third to T (which means that the first and second also form an assignable set). In both cases there would exist more than the two initial assignable subsets.

Let us formally state a definition:

Definition 5.1:

Given a set of jobs X, a set of workers Y and a qualification function $\Phi : X \rightarrow P(Y)$, the set $A \subset X$ is <u>assignable</u> if there is a function $\varphi : A \rightarrow Y$, called an assignment function, which is injective and such that, for each $x \in A$, $\varphi(x) \in \Phi(x)$.

In what follows, we denote by $\mathcal{Q}$ the collection of all assignable sets in a given assignment problem.

We shall show that there exists an assignable set which is optimal. This set will be optimal with respect to the order relation "more important than" defined on X. Its existence is a consequence of Theorem 4.1 and the following result:

Theorem 5.2

Given a set of jobs X, a set of workers Y and a qualification function Φ the collection $\mathcal{Q}$ of the assignable sets is the collection of the independent sets of a matroid on X.

Proof.

Obviously, the empty set is assignable and a subset of an assignable set is also assignable. Therefore, only axiom J3 remains to be verified.

Let $A, A' \in \mathcal{Q}$ and $|A'| = |A| + 1$. We have to prove that there is $x \in A' - A$ such that $A \cup \{x\} \in \mathcal{Q}$. Let φ and φ' be the assignment functions of A and A', respectively. We distinguish two cases:

Case I

Suppose there is an element $x \in A' - A$ such that $\varphi'(x) \notin \varphi(A)$, where $\varphi(A)$ obviously denotes the image of A under φ. We set $B = A \cup \{x\}$ and extend φ to B by setting $\varphi(x) = \varphi'(x)$. The set B is therefore assignable.

Case II

Suppose that $\varphi'(A' - A) \subset \varphi(A)$. Since $|A'| = |A| + 1$ and φ and φ' are injective, we have $|\varphi'(A')| = |\varphi(A)| + 1$. As a consequence, if $|\varphi(A) - \varphi'(A')| = p$, then $|\varphi'(A') - \varphi(A)| = p + 1$ and therefore, there are $p + 1$ elements in $A' \cap A$, denoted $z_{0,1}, \dots, z_{0,p+1}$, whose images under φ' are not in $\varphi(A)$.

Consider now the elements $\varphi(z_{0,1}), \dots, \varphi(z_{0,p+1})$ of $\varphi(A)$. Since $|\varphi(A) - \varphi'(A')| = p$, there are q of these elements, say $\varphi(z_{0,1}), \dots, \varphi(z_{0,q})$ with $q \geq 1$, which are images under φ' of elements in A', say $z_{1,1}, \dots, z_{1,q}$; in symbols, for $r = 1, \dots, q$, $\varphi'(z_{1,r}) = \varphi(z_{0,r})$, or $z_{1,r} = \varphi'^{-1}(\varphi(z_{0,r}))$. If one of these, say $z_{1,r}$, is in $A' - A$, set $x = z_{1,r}$. If all $z_{1,r}$ are in $A' \cap A$, then note first that $z_{1,1}, \dots, z_{1,q}, z_{0,1}, \dots, z_{0,p+1}$ are all distinct: this follows from the fact that their images under φ' are all distinct. Now consider their images under φ, which are all in $\varphi(A)$. Again, because $|\varphi(A) - \varphi'(A')| = p$, there are $q + q'$ of these elements, with $q' \geq 1$, which are images under φ' of elements in A', say $z_{2,1}, \dots, z_{2,q'}, z_{1,1}, \dots, z_{1,q}$; in symbols, for $r = 1, \dots, q'$, $z_{2,r} = \varphi'^{-1}(\varphi(z_{1,r}))$. We repeat this reasoning and, because $A' \cap A$ is finite, we can form a sequence $z_{0,r}, z_{1,r}, \dots, z_{i,r} = x$ where, for $j = 1, \dots, i$, $z_{j,r} = \varphi'^{-1}(\varphi(z_{j-1,r}))$, and x is the first (and only) element of the sequence which belongs to $A' - A$. We set $B = A \cup \{x\}$ and $\varphi'' : B \rightarrow Y$ with $\varphi''(z_{0,r}) = \varphi'(z_{0,r})$, $\varphi''(z_{j,r}) = \varphi(z_{j-1,r})$ for $j = 1, \dots, i$, and $\varphi''(w) = \varphi(w)$ for $w \in A - \{z_{0,r}, z_{1,r}, \dots, z_{i-1,r}\}$. The set B is therefore assignable, which completes the proof of the theorem.

The existence of an optimal assignable set is therefore guaranteed and we can use Kruskal's algorithm to construct it.

For later reference; we mention here a reformulation of this assignment problem which uses (discrete) time slots $Y = \{y_1, \dots, y_m\}$ instead of workers and a time slot allocation function $\Phi : X \rightarrow P(Y)$ instead of a qualification function. More particularly (see Horowitz and Sahni [4]), a set of (discrete) deadlines $d_1, \ldots, d_n$ associated to the jobs $x_1, \ldots, x_n$, respectively, may be given.

6. A CLASS OF FUNCTIONS WHICH GENERATE MATROIDS

What we have seen in the previous sections can be generalized. We need some auxiliary results, the first being a very useful theorem to generate matroids, of which we give an elementary proof.

Theorem 6.1

Let E be a finite set and $f : P(E) \rightarrow Z$ a function whose range is in the set Z of the nonnegative integers and which has the following three properties:

F1) $f(\emptyset) = 0$;

F2) $\forall A,B \in P(E) : A \subset B \Rightarrow f(A) \leq f(B)$;

F3) $\forall A,B \in P(E) : f(A) + f(B) \geq f(A \cup B) + f(A \cap B)$.

Then the subsets X of E such that $\forall Y \subset X : f(Y) \geq |Y|$ form the collection of independent sets of a matroid on E.

Proof.

Axioms J1 and J2 can be immediately verified. Let us verify axiom J3.

Let A,B be subsets of E with $|A| = |B| + 1$ and such that, for any subset Y of A or B, $f(Y) \geq |Y|$. Suppose J3 does not hold for A and B, that means, there is no x in A-B such that $B \cup \{x\}$ fulfills the condition $\forall Y \subset B \cup \{x\} : f(Y) \geq |Y|$. This is equivalent to saying that, for every x in A - B, there is (at least) one set $Y_x \subset B$ such that $f(Y_x \cup \{x\}) < |Y_x \cup \{x\}| = |Y_x| + 1$. By the hypothesis, however, $f(Y_x) \geq |Y|$. By the definition of f, $f(Y_x) \leq f(Y_x \cup \{x\})$. We therefore have $|Y_x| \leq f(Y_x) \leq f(Y_x \cup \{x\}) < |Y_x| + 1$.

which means $f(Y_x) = f(Y_x \cup \{x\}) = |Y_x|$. Now, let $x_1, \ldots, x_t$ be the elements of A - B. Among the sets Y which we can associate, as we just said, to the element x_i, let us choose one with maximum cardinality, denoted Y_i. We then have, for i = 1, ..., t, $f(Y_i) = f(Y_x \cup \{x_i\}) = |Y_x|$. We now claim that $Y_1 = \ldots = Y_t = Y$.

Suppose the contrary, this means, suppose that, with no loss of generality, $Y_1 \neq Y_2$. The definitions of Y_1 and Y_2 and the properties of finite cardinals allow us to write,

$$f(Y_1 \cup Y_2) + f(Y_1 \cap Y_2) \leq f(Y_1) + f(Y_2) = |Y_1| + |Y_2| = |Y_1 \cup Y_2| + |Y_1 \cap Y_2|.$$

But $Y_1 \cup Y_2 \subset B$ and $Y_1 \cap Y_2 \subset B$, hence $f(Y_1 \cup Y_2) \geq |Y_1 \cup Y_2|$ and $f(Y_1 \cap Y_2) \geq |Y_1 \cap Y_2|$. We thus have $f(Y_1 \cup Y_2) = |Y_1 \cup Y_2|$ and $f(Y_1 \cap Y_2) = |Y_1 \cap Y_2|$. On the other hand, $f(Y_1 \cup \{x_1\}) \cup Y_2) + f((Y_1 \cup \{x_1\}) \cap Y_2) = f(Y_1 \cup Y_2 \cup \{x_1\}) + f(Y_1 \cap Y_2) \leq f(Y_1 \cup \{x_1\}) + f(Y_2) = |Y_1| + |Y_2|$, this means, $f(Y_1 \cup Y_2 \cup \{x_1\}) \leq |Y_1| + |Y_2| - |Y_1 \cap Y_2| = |Y_1 \cup Y_2|$. We therefore have $f(Y_1 \cup Y_2) = f(Y_1 \cup Y_2 \cup \{x_1\}) = |Y_1 \cup Y_2|$, which implies $Y_1 \cup Y_2 = Y_1$ otherwise we contradict the maximality of Y_1. Replacing x_1 by x_2, we obtain $Y_1 \cup Y_2 = Y_2$, otherwise we contradict the maximality of Y_2. It is therefore $Y_1 = \ldots = Y_t = Y$, as we claimed.

By the definitions of f and Y,

$$f(Y \cup \{x_1\}) = f(Y \cup \{x_2\}) = f(Y) = |Y|$$

implies now

$$f(Y \cup \{x_1\}) \cup \{x_2\}) + f(Y) \leq |Y| + |Y| ;$$

and, consequently, $f(Y \cup \{x_1\} \cup \{x_2\}) = |Y|$. By induction we obtain

$$f(Y \cup \{x_1\} \cup \{x_2\} \cup \ldots \cup \{x_t\}) = |Y|.$$

Set now $W = A \cap B - Y$. Since $A - B = \{x_1, \ldots, x_t\}$ and $Y \subset B$, we have

$$f(A-W) \leq f(Y \cup \{x_1\} \cup \ldots \cup \{x_t\}) = |Y| \leq |B| - |W| < |A| - |W| = |A-W|,$$

which contradicts our initial hypothesis on A.

Axiom J3 must therefore hold, which completes the proof.

7. APPLICATIONS TO HYPERGRAPHS AND MATROIDS ASSOCIATED TO THEM.

We introduce now the concept of a hypergraph, one of the most important generalizations of the concept of a graph.

Definition 7.1

A hypergraph $H = (Y, \mathcal{E})$ is a system formed by a finite set Y and a collection $\mathcal{E}$ of subsets of Y. The elements of Y are called the vertices of H and the elements of $\mathcal{E}$ are called the hyperedges of H (see Berge [1]).

Graphs with no loops and no multiple edges are hypergraphs whose hyperedges have exactly two vertices; graphs with loops are hypergraphs whose hyperedges have one or two vertices.

We associate to a hypergraph H a class of matroids in the following way:

Let a and b be integers and $a > 0$. Set $\xi_1 = \{A : A \in \xi$ and $|A| \leq -b\}$ and $\xi_2 = \{A : A \in \xi$ and $|A| > -b\}$; let $f : P(\xi_2) \to Z$ be a function such that $f(\emptyset) = 0$ and $f(\beta) = a(|\bigcup_{B \in \beta} B| + b)$ for $\beta \neq \emptyset$ and $\beta \subset \xi_2$. For brevity's sake, let us henceforth denote $|\bigcup_{B \in \beta} B|$ by R. Clearly, R represents the number of elements of Y (or vertices of H) which belong to some hyperedge B of β. Since $\beta \subset \xi_2$, we have $R > -b$; as a consequence, the function f is nonnegative and satisfies conditions F1, F2 and F3 of Theorem 6.1, as is easy to verify. This function defines therefore a matroid on ξ_2 : the independent sets of this matroid are the subsets α of hyperedges such that $\forall \gamma \subset \alpha : f(\gamma) \geq |\gamma|$ where $|\gamma|$ is obviously the number of hyperedges in γ. Note that ξ_1 may be non-empty; in such a case, the matroid we just defined may be extended to ξ if we call a subset η of ξ independent in the extended matroid whenever $\eta \subset \xi_2$ and η is independent in the initial matroid. This extension is a natural one because ξ_1 and ξ_2 are disjoint.

The extended matroid which is associated in this way to the hypergraph $H = (Y, \xi)$ using the integers a and b, will be denoted M(H,a,b).

8. EXAMPLES OF MATROIDS M(H,a,b) WHEN THE HYPERGRAPH H IS A SIMPLE GRAPH.

As we shall see in the next section, hypergraphs and the matroids M(H,a,b) associated to them are very useful for generalizing the assignment problem dealt with in Section 5. They are also interesting because they are the same matroids which arise from some of the so-called matroidal families of graphs; these matroids have recently been extensively investigated (see [3], [9], [10], [11], [12], [14]). For a survey, see [13].

Definition 8.1

A matroidal family of graphs is a non-empty collection Q of finite, connected graphs with the following property : Given an arbitrary graph $H = (Y, \xi)$, the edge sets of the subgraphs of H which are isomorphic to some member of Q are the circuits of a matroid, denoted Q(H), on ξ

The collection Q_1 of all the polygons or cycles of length at least three is a matroidal family: $Q_1(H)$ is the polygon matroid of H, as introduced in Section 2.

Consider now the matroids M(H,a,b) where the hypergraph $H = (Y, \xi)$ is a (simple) graph and b is 0, -1 or -2. The circuits of each one of these matroids are the edge sets of the subgraphs of H which belong to a certain matroidal family.

In fact, Lorea [9] and, independently, Schmidt [11] have proved the following result: - "Let n and s be integers, $n \geq 0$, $-2n \leq s \leq 0$. Let $Q_{n,s}$ be the collection of all graphs $G = (V,E)$ such that: i) $n|V| + s + 1 = |E|$ and $|V| \geq 2$; and: ii) G has no proper subgraph with property (i). Then $Q_{n,s}$ is a matroidal family". On the other hand, the circuits of the matroid M(H,a,b) are the minimal subsets β of ξ such that $f(\beta) = aR + ab < |\beta|$ or, equivalently, the minimal subsets β of ξ such that $aR + ab + 1 = |\beta|$. Recalling the meaning of R, it follows that, for b equal to 0, -1 or -2, the circuits of M(H,a,b) are the circuits of $Q_{a,ab}(H)$.

In particular, for a = 1, M(H,1,-2) is the trivial (uniform) matroid whose only independent set is the empty set, M(H,1,-1) is the polygon matroid and M(H,1,0) is the bicircular matroid. The circuits of the bicircular matroid are the edge sets of the bicircular graphs or bicycles, which are the graphs obtained by a (possibly empty) sequence of edge subdivisions of one of the graphs in figure 8.1.

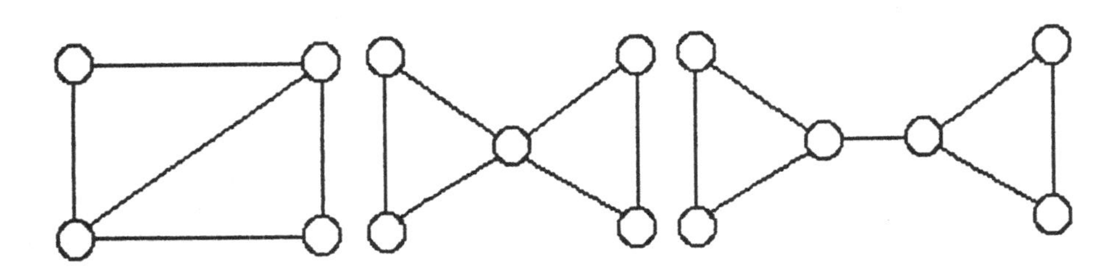

Figure 8.1

9. MATROIDS M(H,a,b) AND GENERALIZED ASSIGNMENT PROBLEMS

Hypergraphs and the matroids M(H,a,b) associated to them allow us to generalize the assignment problem dealt with in Section 5 (see Lorea [7-8]). This generalization becomes more meaningful if we now think in terms of time slots instead of workers.

Consider the hypergraph $H = (Y, \xi)$, let Y be a set of time slots, ξ a set of jobs and, for $y \in Y$ and $A \in \xi$, let $y \in A$ mean that the job A can be assigned to the time slot y, or, in other words, the time slot y is "qualified" to receive the job A. Again, let the jobs be linearly ordered according to a certain criterion for their importance.

In this generalized version of the problem, we suppose that, in any assignment, the number of jobs executed in each time slot is between 0 and a; and each job is either not assigned or assigned to precisely one time slot. Moreover, we consider the possibility of using additional time slots or of becoming unable to use some of the slots we initially have.

Our problem is to find, in this generalized setting, an assignment of jobs to the time slots which is optimal in the sense of Definition 3.1.

We will see that the assignable sets of jobs are the independent sets of the matroids M(H,a,b).

Consider first the case where $b = 0$. Naturally, we say that a set of jobs $\beta \subset \xi$ is assignable when there is a set of time slots $W \subset Y$ and a function $\varphi : \beta \to W$ such that $\forall A \in \beta : \varphi(A) \in A$ and $\forall w \in W : |\{A : \varphi(A) = w\}| \leq a$. In words: φ asigns to each job a time slot "qualified" to receive it and each time slot receives at most a jobs. We then say that the jobs in the set β are assignable to the time slots in the set W. On the other hand, by the definition of the function f which generates M(H,a,b), if β is a set of jobs, then R is the total number of time slots to which al least one of these jobs can be assigned taking into account the allocation possibilities described by the hypergraph H. As a consequence, β is an independent set of M(H,1,0) if and only if the inequality $f(\beta) = R \geq |\beta|$ holds for β and its subsets, which means that, for the set of jobs β or any subset of β, there are enough "qualified" time slots to which the jobs can be assigned. For $b = 0$ and $a = 1$, the problem is therefore the same we dealt with in Section 5, now reformulated in terms of hypergraphs.

Consider now the case where $b < 0$. Let us say here that a set of jobs $\beta \subset \mathcal{E}$ is assignable when there exists, for each subset $Y' \subset Y$ with $|Y'| = |b|$, a set $W \subset Y - Y'$ and a function $\varphi : \beta \to W$ such that, as above, $\forall A \in \beta : \varphi(A) \in A$ and $\forall w \in W : |\{A : \varphi(A) = w\}| \leq a$. This means that the jobs in β can be assigned to time slots qualified to receive them, each slot receiving at most a jobs, and this assignement is possible even if an arbitrary set of $|b|$ slots have meanwhile become unavailable; in other words, we can say that the total time available has been decreased by $|b|$ slot units.

To show that the assignable sets of jobs are the independent sets of M(H,a,b) in this case, recall again the definition of the function f; if β is a set of jobs, then R is still the total number of time slots to which at least one of these jobs can be assigned taking into account the allocation constraints described by H; $R - |b|$ represents this total number of time slots decreased by the maximum number of those among them which might have become unavailable; $a(R - |b|)$ is the number of assuredly available time slots multiplied by the number of jobs each one can receive; and $a(R - |b|) \geq |\beta|$ means that this total is at least equal to the number of jobs to be assigned. If this inequality holds for β and its subsets, then β is assignable to any set of time slots obtained after decreasing the available total time by an arbitrarily chosen set of $|b|$ time slots.

Finally, when $b > 0$, let Y' be disjoint from Y and $|Y'| = b$. A set β is called assignable when there exists W and a function $\varphi : \beta \to W \cup Y'$ such that $\forall A \in \beta : \varphi(A) \in A \cup Y'$ and $\forall w \in W \cup Y' : |\{A : \varphi(A) = w\}| \leq a$. This means that the jobs of the set β can be assigned provided that we can be given b additional time slots, each one allowed to receive any one of the jobs. We leave it as an exercise to the reader to show that the sets assignable under these conditions are also the independent sets of the matroid M(H,a,b).

In terms of workers, this latter case would mean hiring b additional workers each one with qualifications to execute all the jobs and the former case would mean firing any set of $|b|$ workers among those initially available.

10. OPEN PROBLEMS

The area is widely open to further research. Little is known about the matroids M(H,a,b) ingeneral and the matroids generated by matroidal families of graphs, except for a few of them.

We single out two kinds of problems which naturally come to mind. The first is whether we can use matroids M(H,a,b) to deal with a situation where we can hire additional workers but not all of them are qualified to execute all the jobs, or we do fire |b| workers but not any set of |b| workers (this means, we choose the set of |b| workers to be fired according to some criterion so that the set of fired workers is optimal in some sense) : when dealing with assignment of jobs to workers this situation is more realistic. The second kind of problem is whether the matroids yielded by the rich collection of matroidal families of Lorea and Schmidt have any interpretation in terms of (possibly more complex) assignment problems.

REFERENCES

1. **C. Berge**, Graphes et Hypergraphes, Dunod Editeur, Paris 1970.

2. **D. Gale**, Optimal assignments in an ordered set: an application of matroid theory, J. Comb. Theory **4** (1968), 176-180.

3. **R. Halin**, Graphentheorie I, II, Wissenschaftliche Buchgesellschaft Darmstadt, Darmstadt, FRG, 1980, 1981.

4. **E. Horowitz, S. Sahni**, Fundamentals of Computer Algorithms, Computer Science Press Inc., Potomac Maryland 1978.

5. **V. Klee**, The greedy algorithm for finitary and cofinitary matroids, Combinatorics (T.S. Motzkin, ed.), Proc. Symposium in Pure Math., vol. XIX, Amer. Math. Soc. 1971.

6. **J.B. Kruskal**, On the shortest spanning subtree of a graph, Proc. Amer. Math. Soc. **7** (1956), 48-50.

7. **M. Lorea**, Matroides sur les ensembles d'arêtes d'hypergraphes, Cahiers Centre d'Etudes Recherche Operationnelle **20** (1978) nº 2, 127-136.

8. **M. Lorea**, La structure de certains problèmes d'affectation, R.A.I.R.O. Recherche Operationnelle **10** (1976) nº 10, 54-63.

9. **M. Lorea**, On matroidal families, Disc. Math. **28** (1979), 103-106.

10. **L.R. Matthews**, Bicircular matroids, Quart. J. of Math, Oxford (2) **28** (1977), 213-228.

11. **R. Schmidt**, On the existence of uncountably many matroidal families, Disc. Math. **27** (1979) 93-97.

12. **J.M.S. Simões-Pereira**, On matroids on edge sets of graphs with connected subgraphs as circuits-II, Disc. Math. **12** (1975), 55-78.

13. **J.M.S. Simões-Pereira**, Matroidal families of graphs, in Combinatorial Geometries: Advanced Topics (Neil L. White, ed.), Cambridge University Press, to appear.

14. **D.K. Wagner**, Connectivity in bicircular matroids, J. Combin. Theory Ser. B **39** (1985), 308-324.

15. **D. Welsh**, Matroid Theory, Academic Press, 1976.

ALGORITHMS FOR POLYMATROID OPTIMIZATION

Wang Zhemin

Dept. of Statistics & Operations Research
Fudan University, Shanghai, China

1. Introduction

Let $M=(S,\mathcal{I})$ be a matroid where S is the ground set and $\mathcal{I}$ is the family of independent set. For $i \in S$ and $I \in \mathcal{I}$, if $\{i\} \cup I \notin \mathcal{I}$, then there exists a unique circuit in $\{i\} \cup I$, which is called a fundamental circuit w.r.t. i and I. Many algorithms in matroid optimization tie up strongly with fundamental circuits. In this paper, we briefly explain an idea which extends the concept of fundamental circuit to polymatroids so that such algorithms can be generalized to polymatroids, correspondingly.

2. Preliminaries

Let S be a finite set and R_+ the set of non-negative real numbers. We denote by R_+^S the set of vectors with index set S and components in R_+, and for $x \in R_+^S$ and $A \subseteq S$, denote $x(A)=\sum_{i \in A} x_i$. A polymatroid P on S is a non-empty compact set of R_+^S such that

$$P = \{ x : x(A) \leqslant f(A) \text{ for any } A \subseteq S\}$$

for some function $f: 2^S \to R_+$ satisfying the following conditions:

(1) $f(\emptyset) = 0$.

(2) For $A \subset B \subseteq S$, $f(A) \leqslant f(B)$.

(3) For $A,B \in 2^S$, $f(A) + f(B) \geqslant f(A \cup B) + f(A \cap B)$.

Furthermore, P is a matroid if f takes integers and, in addition, satisfies

(4) $f(\{i\}) \leqslant 1$ for all $i \in S$.

A vector x is said to be independent if $x \in P$. x is called a (proper) subvector of y if $x \leqslant y$ $(x < y)$, i.e., for every $i \in S$, $x_i \leqslant y_i$ (and for some $i \in S$, $x_i < y_i$). x is called the maximal independent subvector of y if there does not exist z such that $x < z \leq y$. The polymatroid P has the following properties:

P1. If $x \in P$, $y \in R_+^S$ and $y \leq x$ then $y \in P$.

P2. For any $y \in R_+^S$, all maximal independent subvectors of y are of the same modulus, where the modulus of vector x is defined to be $x(S)$.

A polymatroid can also be defined to be a nonempty compact subset of R_+^S satisfying P1 and P2 [3,8].

3. Fundamental Circuits

The fundamental circuit w.r.t. $i \in S$ and $x \in P$ is defined to be

$$C(i,x) = \begin{cases} \emptyset, & \text{if } \nexists\ A \subseteq S \text{ such that } i \in A \text{ and } x(A)=f(A), \\ \cap\{A : A \subseteq S,\ i \in A \text{ and } x(A)=f(A)\}, & \text{otherwise.} \end{cases}$$

Let

$$m(i,x) = \max\{y_i : y \in P \text{ such that } y_k = x_k \text{ for } k \in S, k \neq i\}$$

and

$$m(i,j,x) = \max\{y_i : y \in P \text{ such that } y_j = 0 \text{ and } y_k = x_k \text{ for all } k \in S \text{ with } k \neq i,j\}.$$

Then we have

Theorem 1. The fundamental circuit $C(i,x)$ can be equivalently defined to be

$$C(i,x) = \begin{cases} 0, & \text{if } m(i,x) > x_i, \\ \{i\} \cup \{j : j \in S \text{ and } m(i,j,x) > x_i\}, & \text{otherwise.} \end{cases}$$

Proof. We first prove that $m(i,x)=x_i$ if and only if there exists $A \subseteq S$ such that $i \in A$ and $x(A)=f(A)$. It is equivalent to say that $m(i,x) > x_i$ if and only if for all $A \subseteq S$ with $i \in A$, $x(A) < f(A)$. For the only if part, suppose $m(i,x) > x_i$. Then, there is $y \in P$ such that $x_i < y_i$ and $x_k = y_k$ for all $k \in S$ with $k \neq i$. Thus, for every $A \subseteq S$ with $i \in A$, we have $x(A) < y(A) \leq f(A)$. For the if part, suppose that for every $A \subseteq S$ with $i \in A$, $x(A) < f(A)$. Denote $a=\min\{f(A)-x(A) : A \subseteq S \text{ and } i \in A\}$. Then $a > 0$. Choose y such that $y_i = x_i + a$ and

$y_k=x_k$ for all $k \in S$ with $k=i$. For any $A \subseteq S$, it $i \in A$ then $y(A)=x(A) \leq f(A)$, and if $i \in A$ then $y(A)=x(A)+a \leq f(A)$. Thus, $y \in P$ and hence $m(i,x)>x_i$.

Similarly, we can prove that $m(i,j,x)=x_i$ if and only if there exists $A \subseteq S$ such that $i \in A$, $j \notin A$ and $x(A)=f(A)$. This completes our proof. □

Theorem 1 tells us that the membership of $C(i,x)$ can be determined by computing $m(i,x)$ and $m(i,j,x)$. Such computation will be oracle in our algorithms.

4. Main Results

The following results are obtained through studying on fundamental circuits.

<u>Theorem</u> 2. Let $y \in (R_+ \cup \{+\infty\})^S$ and $c \in R^S$. Then x^* is the optimal solution for

$$\max\{cx : x \text{ is over all maximal independent subvectors of } y\}$$

if and only if for any $i \in S$, if $x_i^* < a_i$ then for every $j \in C(i,x^*)$ with $j=i$, $c_i \leq c_j$.

<u>Theorem</u> 3. Let $c \in R^S$. Then x^* is the optimal solution for

$$\max\{cx : x \in P\}$$

if and only if the following conditions hold,

(a) if $i \in S$ and $c_i < 0$ then $x_i^* = 0$,

(b) if $i \in S$ and $c_i > 0$, then $C(i,x^*) \neq 0$ and for any $j \in C(i,x^*)$ with $j \neq i$, $c_i \leq c_j$.

Several results in [1,3,2,7] can be derived from the above theorems. Based on these theorems and those results, we designed several algorithms for optimization problems in Theorem 1 and Theorem 2, with the complexity of using oracle at most $O(|S|^k)$ times where k=1,2 or 3.

References

1. F.D. Dunstan and D.J.A. Welsh, A greedy algorithm for solution of certain class of linear programming, Math. Program. 5(1973) 338-353.

2. R.E. Bixby, W.H. Cunningham and D.M. Topkis, The partial

order of polymatroid extreme point, Math. of OR 10 (1985) 367-378.

3. J. Edmonds, Submodular functions, matroids and certain polyhedra, Proc.Int.Conf.on Combinatorics (Calgary), Gordor and Breach (New York) (1970) 69-87.
4. J. Edmonds, Matroids and the greedy algorithm, Math. Pragram. 1 (1971) 127-136.
5. A. Frank, Generalized polymatroid, Proc. of the sixth Hungarian Colloquim, 1981.
6. E.L. Lawler, Matroid intersection algorithms, Math. Program. 9 (1975) 31-56.
7. D.M. Topkis, Adjacency on polymatroid, Math. Program. 30 (1984) 229-237.
8. D.J.A. Welsh, Matroid Theory, Academic Press, London, 1976.

FREE PARTIALLY COMMUTATIVE GROUPS*

C. WRATHALL

Department of Mathematics
University of California
Santa Barbara, California 93106
U.S.A.

1. INTRODUCTION

It may happen that some of the operations available for computations are independent. Consider, for example, the following three operations:

(a) Increment register 1.

(b) Increment register 2.

(c) Assign the product of registers 1 and 2 to register 1.

Clearly a sequence of operations (a) and (b) can be performed in any order (or even in parallel): whatever order is used, the final contents of the registers will be the same. However, operation (c) does not have this freedom since (depending on the contents of the registers) "perform (b), then (c)" might not have the same final result as "perform (c), then (b)". We may say that (a) and (b) are independent or commuting operations, but that (c) is independent of neither (a) nor (b).

In such a setting of independent operations, questions arise concerning sequences of operations, for example, the "Word Problem": are two given sequences equivalent? A sequence of

---

* This paper is based on a lecture given at the International Symposium on Combinatorial Optimization, Nankai Institute of Mathematics, Tianjin, People's Republic of China in August 1988. Preparation of the paper was supported in part by the National Science Foundation under Grant No. CCR-8706976.

operations can be viewed as a string of symbols that may be rewritten by exchanging adjacent symbols when they represent independent operations; the specification of a particular system states that symbols representing independent operations should commute. From another point of view, an observer may make a sequential record (or "trace") of a computation in which some operations are performed in parallel, and specifying that some symbols commute recaptures the independence of those parallel operations. The Word Problem for such a system (from either point of view) can be restated as asking whether one given sequence can be transformed into the other by a (finite) number of exchanges of adjacent commuting symbols.

Questions about sequences of operations in this context have been studied formally under the names of "commutation monoids", "free partially commutative monoids", and "traces". (See, for example, [22, 26, 27].) Recent attention has been given to commutation monoids due to their use in modeling problems of concurrency control in databases [14] and their connections to Petri nets [1]. They were also studied earlier from a combinatorial point of view by Cartier and Foata [6].

The set of finite-length sequences (that is, strings) forms a monoid: juxtaposition of sequences (concatenation of strings) is an associative "multiplication" operation and the empty sequence or string serves as an identity for the multiplication. In a free monoid, no further relationship among the strings is assumed; in a commutation monoid, some of the letters making up the strings may be independent, that is, may commute.

The Word Problem for commutation monoids has been shown to be solvable in linear time as a function of the length of the input strings [5, 26]. When none of the letters commute (and so the strings lie in a free monoid), the problem is easily solved by comparing the strings symbol-by-symbol. In the more general context, the sequences can be converted to certain "normal forms" and then compared, or they can be projected to tuples of strings that are then compared.

Suppose we require in addition that every operation a possess an inverse operation $\overline{a}$. The first operation above, for example, has the inverse operation "Decrement register 1". The resulting structure may be modelled by a free partially commutative group: the elements are represented by strings of letters, every letter (and so every element) has an inverse, and some of the letters are specified to commute. The term "free partially commutative group" is used here (rather than "commutation group") to emphasize that any choice of commuting letters can be made. Free partially commutative groups thus fall in a range between free groups (in which no letters commute) and free Abelian groups (in which all letters commute).

It is shown here that, for a fixed free partially commutative group G, there are linear-time algorithms to solve the Word Problem and the Conjugacy Problem for G.

A free partially commutative group can be presented as a monoid by adding formal inverses for the generating letters, and using the relations to express both the properties of the inverses and the partial commutativity. If partial commutativity is specified completely in the presentation, then the monoid presentation is preperfect as a Thue system (Theorem 4). As a consequence, for a finitely presented free partially commutative group there is a linear-time algorithm to produce a "projected" normal form of a given string, where the projected form is a tuple of strings over certain subsets of the original alphabet. From this projected normal form, one can construct a shortest string equivalent to the given string; also, a pair of projected normal forms can be easily compared, to solve the Word Problem for the group in linear time (Theorem 6). Another linear-time algorithm, due to Duboc [13], is also described here.

Elements g_1 and g_2 of a group are called "conjugate" elements if there is some element h such that $h \cdot g_1 \cdot h^{-1}$ is the same element as g_2. Elements m_1 and m_2 of a monoid are conjugate if there is some element n such that $n \cdot m_1$ is the same element as $m_2 \cdot n$. In general, this relation on monoids is not an equivalence relation; however, it is an equivalence relation for commutation monoids, including free monoids. Whether elements of a free partially commutative group G are conjugate can be tested in linear time (Theorem 10): the Conjugacy Problem for G reduces to a problem of conjugacy in a commutation monoid M, and conjugacy problems in commutation monoids can be solved in linear time (by a reduction to pattern-matching questions [19]).

The schemes described here for solving the Word and Conjugacy Problems will also serve when only some of the generating letters have inverses, that is, when only some of operations represented by the symbols are invertible.

The results presented here draw on two lines of research in theoretical computer science: first, they are an application of work on the algebraic properties of commutation monoids [9, 8, 12], especially the device of translating between a commutation monoid and a certain product of free monoids; and second, they are a special case of questions about rewriting systems. The method used to show the preperfect property (which underlies the other results) is based on recasting the question and appealing to well-known facts about abstract rewriting systems.

2. PRELIMINARY DEFINITIONS AND NOTATION

The notation and terminology used here for sets and free monoids follows that in the book of Lothaire [20]. The basic notions of computation on strings can be found in the texts by Hopcroft and Ullman [15], or Aho, Hopcroft and Ullman [2].

For a finite set of letters (alphabet) A, A^* denotes the free monoid generated by A. The empty string, the identity element of the free monoid, is denoted by 1. The length of string x, denoted by $|x|$, is the number of occurrences of symbols in x; in particular, $|1| = 0$. For a string $x \in A^*$ and letter $a \in A$, $|x|_a$ denotes the number of occurrences of the letter a in x. The set $\mathrm{alph}(x) = \{ a \in A : |x|_a > 0 \}$ is the set of letters that occur in the string x.

A Thue system T on an alphabet Σ is a set of pairs of strings, $T \subseteq \Sigma^* \times \Sigma^*$. For $u, v \in \Sigma^*$, write $u \leftrightarrow v$ if there is some rule $(x,y) \in T$ such that, for some r and s, either $u = rxs$ and $v = rys$, or $u = rys$ and $v = rxs$. The congruence $\leftarrow\!*\!\rightarrow$ generated by T is the reflexive and transitive closure of $\leftrightarrow$; the congruence class of $u \in \Sigma^*$ is $[u] = \{ v : u \leftarrow\!*\!\rightarrow v \}$. The Thue system T is a presentation of $\Sigma^*/\leftarrow\!*\!\rightarrow$, the quotient monoid of Σ^* by $\leftarrow\!*\!\rightarrow$. The reader may refer to the paper by Book [4] or the monograph by Jantzen [17] for a survey of Thue systems.

If $u \leftrightarrow v$, then write $u \;|\!-\!|\; v$ if $|u| = |v|$, $u \longrightarrow v$ if $|u| > |v|$, and $u \;|\!\!\longrightarrow v$ if $|u| \geq |v|$. Let $|\!-\!*\!-\!|$, $-\!*\!\rightarrow$ and $|\!-\!*\!\rightarrow$ denote the reflexive, transitive closures of $|\!-\!|$, $\longrightarrow$ and $|\!\!\longrightarrow$, respectively.

Thue systems are a particular type of rewriting systems, ones that operate on strings, and some of their properties reflect those of abstract rewriting systems. Huet [16] has given a survey, with historical remarks, of these abstract systems. Suppose U is a set and $\Rightarrow$ is any binary relation on U, called a "reduction" relation and written as an infix symbol: $u \Rightarrow v$. Let $\Longleftrightarrow$ denote the symmetric closure of $\Rightarrow$, and $\Rightarrow^*$ and $\Leftarrow\!*\!\Rightarrow$ the reflexive-transitive closures of $\Rightarrow$ and $\Longleftrightarrow$. The relation $\Rightarrow$ is called:

(i) <u>confluent</u> if $u \Rightarrow^* v_1$ and $u \Rightarrow^* v_2$ imply there is some w such that $v_1 \Rightarrow^* w$ and $v_2 \Rightarrow^* w$;

(ii) <u>locally confluent</u> if $u \Rightarrow v_1$ and $u \Rightarrow v_2$ imply there is some w such that $v_1 \Rightarrow^* w$ and $v_2 \Rightarrow^* w$;

(iii) <u>Noetherian</u> if there is no infinite sequence $u_1 \Rightarrow u_2 \Rightarrow \ldots$.

A reduction relation that is Noetherian and locally confluent is confluent, and it also has the "Church-Rosser" property: if $u \Leftarrow^* \Rightarrow v$ then there is some w such that $u \Rightarrow^* w$ and $v \Rightarrow^* w$.

3. FREE PARTIALLY COMMUTATIVE GROUPS

Suppose that Σ_0 is an alphabet and θ_0 is a partial commutativity relation on Σ_0; that is, θ_0 is a symmetric and irreflexive binary relation. The free partially commutative group $G(\theta_0)$ is the group generated by the letters in Σ_0 in which the defining relations are "$ab = ba$" for each pair (a,b) in θ_0.

For the present purpose, it is convenient to use a monoid presentation for $G(\theta_0)$. Let $\Sigma_1 = \{\overline{a} : a \in \Sigma_0\}$ be a set of formal inverses for the letters in Σ_0, and $\Sigma = \Sigma_0 \cup \Sigma_1$. Let $\theta \subseteq \Sigma \times \Sigma$ be the extension of θ_0 to Σ:

$$\theta = \{(a,b), (\overline{a},b), (a,\overline{b}), (\overline{a},\overline{b}) : (a,b) \in \theta_0\}.$$

Let T_0 be the following Thue system on Σ:

$$T_0 = \{ (a\overline{a}, 1), (\overline{a}a, 1) : a \in \Sigma_0\} \cup \{ (cd, dc) : (c,d) \in \theta \}.$$

The Thue system T_0 presents the free partially commutative group $G(\theta_0)$ and, in a sense, it is the "natural" presentation of $G(\theta_0)$ as a monoid. In this context, $\longrightarrow$ is a single free-reduction step and $|-|$ is a step that rearranges a pair of letters according to the (extended) partial commutativity relation. Let $\equiv_G$ denote the congruence generated by the Thue system T_0, so that $G(\theta_0) = \Sigma^*/\equiv_G$. For strings x and y, $x \equiv_G y$ exactly when y can be obtained from x by a finite sequence of commutations (according to θ) and insertions and deletions of inverse pairs.

For $x = a_1 \cdots a_n$, $a_i \in \Sigma$, x^{-1} denotes the string $\overline{a_n} \cdots \overline{a_1}$, which represents the inverse of x in $G(\theta_0)$. (If letter d is $\overline{a}$, then $\overline{d} = a$.)

If θ_0 is empty, then $G(\theta_0)$ is just the free group on the generators Σ_0, and if θ_0 contains every pair of distinct letters, then $G(\theta_0)$ is the free Abelian group on Σ_0. Between these two extremes any choice of pairs of commuting letters in Σ_0 can be made, although (as a consequence of the first set of rules) the letters a and $\overline{a}$ will always commute.

Some of the rules in the second set specifying T_0 may be redundant. For example, if $(a,b) \in \theta_0$ then both (ab, ba) and $(a\overline{b}, \overline{b}a)$ are included, but the second rule is can be derived from the first is included, since $a\overline{b} \leftarrow \overline{b}ba\overline{b} \; |\!-\!| \; \overline{b}ab\overline{b} \longrightarrow \overline{b}a$. This redundancy is necessary if the presentation is to have the "preperfect" property given in Theorem 4.

The commutation monoid corresponding to $G(\theta_0)$ is the monoid $M(\theta) = \Sigma^*/|\!-\!*\!-\!|$ presented by the Thue system $\{ (cd, dc) : (c,d) \in \theta \}$. Thus, strings represent the same element of $M(\theta)$ if one can be transformed into the other by use of commutation rules only. Strings congruent in M have the same length and, indeed, the same number of each letter in Σ.

Strings x and y are <u>independent</u> if $\text{alph}(x) \times \text{alph}(y) \subseteq \theta$; that is, if every letter in x commutes with every letter in y (which implies that $xy \; |\!-\!*\!-\!| \; yx$). Because θ_0 and hence θ are irreflexive relations, independent strings have disjoint alphabets; note also that a string is independent of $a \in \Sigma_0$ if and only if it is independent of $\overline{a}$.

A very useful representation of commutation monoids is as products of certain free monoids. The alphabets for the free monoids can be found by collecting together noncommuting letters, as follows.

Let $A_1, \ldots, A_N$ be a collection of subsets of Σ that cover Σ and have the following properties: for all $a, b \in \Sigma$

(i) for all i, $a \in A_i$ iff $\overline{a} \in A_i$;

(ii) if $(a,b) \in \overline{\theta}$, then, for some j, A_j contains both a and b;

(iii) if, for some i, A_i contains both a and b, then $(a,b) \in \overline{\theta}$.

For example, the collection might consist of each four-element set $\{a, \overline{a}, b, \overline{b}\}$ for $(a,b) \in \overline{\theta_0}$ $(a \neq b$ in $\Sigma_0)$, together with the two-element sets $\{c, \overline{c}\}$ for those letters c that commute with all the other letters. Another choice is to take the collection of "maximal cliques" (i.e., maximally non-commuting sets of letters) of $\overline{\theta_0}$, adding the appropriate inverses to each one.

For each i, let $\pi_i : \Sigma^* \longrightarrow A_i^*$ be the projection of Σ onto A_i, that is, the homomorphism determined by defining $\pi_i(a) = a$ for $a \in A_i$ and $\pi_i(a) = 1$ for $a \notin A_i$. Property (ii), above, ensures that, for any string x and letter a, if $\pi_i(x) = 1$ whenever $a \in A_i$, then x is independent of a.

Let $\Pi : \Sigma^* \longrightarrow A_1^* \times \ldots \times A_N^*$ be the function defined by $\Pi(w) = (\pi_1(w), \ldots, \pi_p(w))$. The set $A_1^* \times \ldots \times A_N^*$ forms a monoid under componentwise concatenation, and Π is a monoid homomorphism. Since the length-preserving rules of T_0 express a partial commutativity relation on Σ, we have the following correspondence between the monoid $M(\theta)$ and the product $A_1^* \times \ldots \times A_N^*$.

Lemma 1 [9, 12]. For all $u, v \in \Sigma^*$, $u \mid\!\overset{*}{-}\!\mid v$ if and only if $\Pi(u) = \Pi(v)$.

A Thue system is <u>preperfect</u> if whenever strings u and v are congruent there is some string w such that $u \mid\!\overset{*}{\longrightarrow} w$ and $v \mid\!\overset{*}{\longrightarrow} w$. An equivalent condition [7] is that if $u \overset{*}{\longleftarrow}\!\mid z \mid\!\overset{*}{\longrightarrow} v$ then, for some w, $u \mid\!\overset{*}{\longrightarrow} w$ and $v \mid\!\overset{*}{\longrightarrow} w$. In this context, preperfect means that the relation $\longrightarrow / \mid\!\overset{*}{-}\!\mid$ is confluent on $M(\theta)$.

A string is <u>minimal</u> relative to a Thue system if it is a shortest string in its congruence class. For a preperfect system, congruent minimal strings are in fact "level-equivalent", that is, if u and v are minimal and $u \overset{*}{\longleftrightarrow} v$ then $u \mid\!\overset{*}{-}\!\mid v$.

For abstract rewriting systems, Jouannaud and Kirchner [18] have studied Church-Rosser property modulo an equivalence relation. Their definition "R is R/E Church-Rosser modulo E" becomes the preperfect property when applied to Thue systems: in that case R is the set of reductions (length-reducing rules) of the system and E is the equivalence relation $|{-}^*{-}|$. Their Theorem 5 then states that a local confluence property modulo E is sufficient for the preperfect property, that is, that the Thue system is preperfect if $u \;|{-}^*{-}| \leftarrow |{-}^*{-}| \rightarrow v$ implies $u \;|{-}^*{\rightarrow} \;{\leftarrow}^*{-}|\; v$.

The route taken here for showing that T_0 is preperfect is by way of a certain reduction relation on tuples of strings: T_0 is preperfect because that reduction relation is confluent. The representation of elements of $G(\theta_0)$ using tuples of strings builds on that given in Lemma 1; it will be used for computing normal forms in $G(\theta_0)$, and for solving the Word and Conjugacy Problems.

Definition. For tuples $s, t \in A_1{}^* \times \ldots \times A_N{}^*$, s <u>reduces to</u> t in one step, written $s \Rightarrow t$, if, for some $a \in \Sigma$ and some $k \geq 0$, for each i,

(i) if $a \notin A_i$ then $t_i = s_i$; and

(ii) if $a \in A_i$ then $s_i = u_i a \overline{a} v_i$, $t_i = u_i v_i$ and $|u_i|_a = k$.

In this definition, the restriction that the deleted pair occur "in the same position" in each component is necessary for the desired correspondence with T_0; that is, performing free reduction in the components independently may lead to incorrect results. Consider, for example, $\Sigma_0 = \{a, b, c\}$ and $\theta_0 = \{ (a,b), (b,a) \}$, with $A_1 = \{a, \overline{a}, c, \overline{c}\}$ and $A_2 = \{b, \overline{b}, c, \overline{c}\}$. The string $w = \overline{c}acb\overline{c}\,\overline{a}c\overline{b}$ is minimal, and, in particular, is not congruent to the empty string, but $\pi_1(w) = \overline{c}ac\overline{c}\,\overline{a}c \;{-}^*{\rightarrow}\; 1$ and $\pi_2(w) = \overline{c}cb\overline{c}c\overline{b} \;{-}^*{\rightarrow}\; 1$.

The following lemmas give the connection between the reduction relation $\Rightarrow$ and the Thue system T_0.

Lemma 2 [28]
(i) For all $u \in \Sigma^*$ and $\mathbf{x} \in A_1{}^* \times \ldots \times A_N{}^*$, if $\Pi(u) \Rightarrow \mathbf{x}$ then there are strings u' and v such that $u \mid\!-^*-\!\mid u' \longrightarrow v$ and $\mathbf{x} = \Pi(v)$.
(ii) For all $\mathbf{r}, \mathbf{s} \in A_1{}^* \times \ldots \times A_N{}^*$ and $u \in \Sigma^*$, if $\mathbf{r} \Leftarrow \Pi(u) \Rightarrow \mathbf{s}$ then either $\mathbf{r} = \mathbf{s}$ or, for some $\mathbf{t}$, $\mathbf{r} \Rightarrow \mathbf{t} \Leftarrow \mathbf{s}$.

Lemma 3. For all $u, v \in \Sigma^*$, $u \mid\!-^*\!\rightarrow v$ if and only if $\Pi(u) \Rightarrow^* \Pi(v)$. Hence, T_0 is preperfect if and only if $\Rightarrow$ is confluent on $\Pi(\Sigma^*)$.

Proof. It is straightforward to verify from the definitions that if $u \longrightarrow v$ then $\Pi(u) \Rightarrow \Pi(v)$, and the implication from left to right of the first statement then follows using induction (on the number of steps in the derivation from u to v) and Lemma 1. The other direction can be proved by induction as well, using Lemma 2(i) and Lemma 1. It is then easy to see that strings u and v are congruent mod T_0 if and only if $\Pi(u) \Leftarrow^*\Rightarrow \Pi(v)$.

Suppose that T_0 is preperfect, and consider any tuples $\mathbf{r}, \mathbf{s}, \mathbf{t} \in \Pi(\Sigma^*)$, with $\mathbf{r} = \Pi(x)$, $\mathbf{s} = \Pi(y)$ and $\mathbf{t} = \Pi(z)$. If $\mathbf{t} \Rightarrow^* \mathbf{r}$ and $\mathbf{t} \Rightarrow^* \mathbf{s}$ then $x \equiv_G y$ so (since T_0 is preperfect) there is some string w such that $x \mid\!-^*\!\rightarrow w$ and $y \mid\!-^*\!\rightarrow w$, and hence $\mathbf{r} = \Pi(x) \Rightarrow^* \Pi(w)$ and $\mathbf{s} = \Pi(y) \Rightarrow^* \Pi(w)$; thus, $\Rightarrow$ is confluent on $\Pi(\Sigma^*)$. On the other hand, suppose that $\Rightarrow$ is confluent on $\Pi(\Sigma^*)$, and strings x and y satisfy $x \equiv_G y$. Then $\Pi(x) \Leftarrow^*\Rightarrow \Pi(y)$, so (using the "Church-Rosser" formulation of the confluence property), there is some $\mathbf{r} \in \Pi(\Sigma^*)$ such that $\Pi(x) \Rightarrow^* \mathbf{r} \; {}^*\!\Leftarrow \Pi(y)$. Taking w to be any string such that $\mathbf{r} = \Pi(w)$, we then have $x \mid\!-^*\!\rightarrow w$ and $y \mid\!-^*\!\rightarrow w$, and hence T_0 is preperfect. ■

The reduction relation $\Rightarrow$ on tuples of strings is a Noetherian relation, since the total length of the strings decreases with each reduction step. From Lemma 2, that relation is also locally confluent on the set $\Pi(\Sigma^*)$, and hence it is confluent. From the connection given in the previous lemma, we can therefore conclude that the Thue system T_0 is preperfect, as stated in the following theorem.

Theorem 4. For a partial commutativity relation θ_0, the Thue system

$$\{ (a\overline{a}, 1), (\overline{a}a, 1) : a \in \Sigma_0 \} \cup \{ (cd, dc) : (c, d) \in \theta \}$$

on alphabet Σ is preperfect.

As a consequence of the preperfect property, if x and y represent the same element of $G(\theta_0)$ and x is minimal then $y \mid\!-^*\!\to x$. Also, if both x and y are minimal and $x \equiv_G y$, then $y \mid\!-^*\!-\!\mid x$, so that x and y represent the same element of $M(\theta)$. [13, Prop. 2.4.7]

A Thue system is <u>almost-confluent</u> if u congruent to v implies that, for some x and y, $u -^*\!\to x \mid\!-^*\!-\!\mid y \leftarrow^*\!- v$. This is a stricter condition than preperfect, since all the reductions must be performed first, rather than being mixed with length-preserving rules. It is decidable whether a (finite) Thue system is almost-confluent [25], but undecidable whether it is preperfect [23] even when the system has a very simple form [24].

The Thue system T_0 is not almost-confluent (unless θ_0 is empty): if a and b are two positive letters that commute (i.e., $(a,b) \in \theta_0$), then $ab\overline{a}$ is congruent to b, but no length-decreasing rule applies to $ab\overline{a}$ since b is neither a nor $\overline{a}$. Hence the relation $ab\overline{a} -^*\!\to x \mid\!-^*\!-\!\mid y \leftarrow^*\!- b$ does not hold for any x and y.

The abstract property that Huet [16] termed "confluence (of a reduction relation) modulo equivalence" becomes the almost-confluent property when applied to Thue systems.

4. Normal Forms and the Word Problem

The Word Problem for a free group can be efficiently solved: each string has a normal form (its freely reduced form) that can be obtained by cancelling all possible inverse pairs $a\bar{a}$ that appear in it, and strings are equivalent in the free group exactly when their normal forms are identical. Based on Lemma 3 and Theorem 4, a "synchronous" extension of that linear-time algorithm will serve for free partially commutative groups. A linear-time algorithm is also known for the word problem for almost-confluent Thue systems in which the length-preserving rules express (only) a partial commutativity relation [5], but that algorithm cannot be applied directly here.

Theorem 5. For each finitely generated free partially commutative group $G(\theta_0)$, there is a linear-time algorithm to find a minimal string congruent to a given string.

Proof. Recall that the function Π maps strings in Σ to tuples of strings in $A_1{}^* \times \ldots \times A_N{}^*$. The construction, given a string w, of a shortest string in $[\dot{w}]$ has two steps, reduction followed by reconstruction. The first step yields the "projected normal form" of w, that is, it yields $\Pi(w_0)$ where w_0 is a minimal string congruent to w. (Since T_0 is preperfect, all the minimal strings in a congruence class are level-equivalent and so, from Lemma 1, have the same image under Π.)

The reduction procedure uses N pushdown stores (initially empty) and operates as follows.

Procedure 1. Read an input symbol, say $a \in \Sigma$. Is $\bar{a}$ the symbol on top of the ith pushdown store whenever a belongs to A_i? If so, erase all those symbols $\bar{a}$ and go on to the next input symbol. If not, write the symbol a onto the top of each store i for which $a \in A_i$ and go on.

This procedure clearly processes an input string w in $|w|$ steps, and produces some tuple $R(w) = (x_1, \ldots, x_N)$ of strings on the pushdown stores (where the top of the store is at the right). A straightforward induction can be used to verify that $R(w)$ is irreducible and $\Pi(w) \Rightarrow^* R(w)$.

Since $\Pi(w) \Rightarrow^* R(w)$, there is (from Lemma 2(i)) some string whose image under Π is $R(w)$; such a string can be printed from right to left by starting with $R(w)$ on the pushdown stores and essentially reversing Procedure 1.

Procedure 2. Examine the pushdown stores to find a letter $b \in \Sigma$ with the property that b is on the top of the ith pushdown store whenever $b \in A_i$. Remove that occurrence of b from each such pushdown store, and print b as the next output letter. Continue until all the pushdown stores are emptied.
If there is more than one candidate for the letter to be printed, any one can be chosen, or a fixed precedence used.

Procedure 2 will eventually empty all the pushdown stores because $R(w)$ is in $\Pi(\Sigma^*)$: if $\Pi(u) = t\,\Pi(b)$ then there is some v such that $\Pi(v) = t$ and $u \ |\!-\!*\!-\!|\ vb$. Let $r(w)$ be the string given as output by Procedure 2 when run on $R(w)$. The time taken is $|r(w)|$, and it is easy to argue that $R(w) = \Pi(r(w))$.

From $\Pi(w) \Rightarrow^* R(w) = \Pi(r(w))$, it follows that $w \ |\!-\!*\!\rightarrow r(w)$. Moreover, $r(w)$ is a shortest string congruent to w: if v were a shorter string, then (since T_0 is preperfect) there would be some string z such that $v \ |\!-\!*\!\rightarrow z \leftarrow\!*\!-\!|\ r(w)$, so $\Pi(v) \Rightarrow^* \Pi(z) \ {}^*\!\Leftarrow \Pi(r(w)) = R(w)$. But $R(w)$ is irreducible, so $\Pi(v) \Rightarrow^* \Pi(r(w))$ with $|v| < |r(w)|$, a contradiction. ■

As an example, consider the alphabet $\{ a, b, c \}$ in which a and b commute but c commutes with neither a nor b; we may take $A_1 = \{ a, \bar{a}, c, \bar{c} \}$ and $A_2 = \{ b, \bar{b}, c, \bar{c} \}$. For the string $w = \bar{c}ab\bar{a}c$, the stacks that result are ($\bar{c}c$, $\bar{c}bc$). A minimal string congruent to w is $y = \bar{c}bc$, and $\pi_1(y) = \bar{c}c$, $\pi_2(y) = \bar{c}bc$.

To solve the Word Problem in $G(\theta_0)$, the algorithm in Theorem 5 can be used, but reconstruction is not necessary: $w_1 \equiv_G w_2$ if and only if their projected normal forms $R(w_1)$ and $R(w_2)$ are identical. Since those two tuples of strings can be constructed and compared in time linear in $|w_1| + |w_2|$, the Word Problem can be solved in linear time.

Theorem 6. For any finitely generated free partially commutative group $G(\theta_0)$, there is a linear-time algorithm for the word problem in the group.

Duboc [13, Section 2.4.2] has used another scheme for reducing a given string to one that is congruent to it and minimal; it is a generalization of one used by Perrin [26] for finding normal forms in commutation monoids. The procedure uses one stack for each letter in Σ_0 and each stack can contain a string over an alphabet of three letters, say, f, g, and $\overline{g}$. Beginning with all the stacks empty, it produces a collection of stacks $S(x) = (S_a(x) : a \in \Sigma_0)$ by reading an input string x from right to left, as follows (here, $\overline{\theta_0}(a) = \{ b \in \Sigma_0 : (a,b) \notin \theta_0 \}$):

When the next letter is $a \in \Sigma_0$:

if the top of S_a is $\overline{g}$

then pop S_a and pop S_b for all $b \in \overline{\theta_0}(a)$;

else push g on S_a and push f on S_b for all $b \in \overline{\theta_0}(a)$.

When the next letter is $\overline{a} \in \Sigma_1$:

if the top of S_a is g

then pop S_a and pop S_b for all $b \in \overline{\theta_0}(a)$;

else push $\overline{g}$ on S_a and push f on S_b for all $b \in \overline{\theta_0}(a)$.

As an example, again suppose $\Sigma_0 = \{a, b, c\}$ where the letters a and b commute but c commutes with neither. For $x = \overline{c}ab\overline{a}c$ (for which a normal form is $\overline{c}b\overline{c}$), the resulting stacks are $S_a(x) = ff$, $S_b(x) = fgf$, and $S_c(x) = gf\overline{g}$.

If x is minimal then no popping will take place while $S(x)$ is being calculated. The collection of stacks $S(x)$ can be reconstructed into a minimal word congruent to x, and $S(x) = S(y)$ if and only if $x \equiv_G y$.

Although these algorithms operate in linear time, the size of the alphabet contributes to the time bound. The uniform Word Problem, in which the input includes both a partially commutative alphabet (Σ_0, θ_0) and two strings, might not have a linear-time solution, but is certainly a tractable problem.

5. The Conjugacy Problem

In a free group, two elements are conjugate if and only if their "cyclic reductions" are conjugate in the corresponding free monoid, where the cyclic reduction of a string is found "by first freely reducing it and then cancelling first and last symbols (if necessary)" [21, p. 36]. A generalization of this fact holds for free partially commutative groups and monoids: strings x and y are conjugate in a free partially commutative group $G(\theta_0)$ if and only if their "cores" $\sigma(x)$ and $\sigma(y)$ are conjugate in the corresponding commutation monoid $M(\theta)$. The core of a string can be found in linear time (Lemma 8) and the conjugacy problem in commutation monoids can be solved in linear time [19]; hence the conjugacy problem in free partially commutative groups can be solved in linear time.

A string in a free group is said to be cyclically reduced if it is freely reduced and does not begin with a letter and end with the inverse of that letter. "Cyclic minimality" is a generalization of this notion to free partially commutative groups. The core of a string is then its inner cyclically-minimal part.

Definition. For $x \in \Sigma^*$, let $F(x) = \{ a \in \Sigma : \text{for some } y, x \mid\!-^*\!-\!\mid ay \}$ and $L(x) = \{ a \in \Sigma : \text{for some } y, x \mid\!-^*\!-\!\mid ya \}$. A string x is <u>cyclically minimal</u> if it is minimal and there is no letter $a \in \Sigma$ such that $a \in F(x)$ and $\overline{a} \in L(x)$, that is, if there is no $y \in \Sigma^*$ and $a \in \Sigma$ such that $x \mid\!-^*\!-\!\mid ay\overline{a}$.

Note that in this definition only length-preserving rules are to be used in transforming x to $ay\overline{a}$.

Definition. For a string $x \in \Sigma^*$, let z be any minimal string congruent to x, and let $u, y \in \Sigma^*$ be any strings such that $z \mid\!-^*\!-\!\mid uyu^{-1}$ and y is cyclically minimal. Then y is a <u>core</u> of x. Let $\sigma(x)$ denote any core of x.

A string is cyclically minimal exactly when its square is minimal; the longer definition is used here in order to make it plain how to compute the core of a string from its reduced projected form. The definition allows different choices for the core of a string, but, from the following lemma, we see that all cores of x represent the same element of the monoid $M(\theta)$.

Lemma 7. If strings uyu^{-1} and vzv^{-1} are congruent in G, both are minimal, and y and z are cyclically minimal, then $y \mid\!-^*\!-\!\mid z$.

Proof. A property of commutation monoids (whether or not of this special form) is that, for each letter $a \in \Sigma$, if $ar \mid\!-^*\!-\!\mid s$, then there exist p and q such that $s = paq$, p is independent of a, and $r \mid\!-^*\!-\!\mid pq$. (This can be easily proved by induction on the number of "exchanging" steps.)

Suppose now that $uyu^{-1} \equiv_G vzv^{-1}$ where uyu^{-1} and vzv^{-1} are minimal and y and z are cyclically minimal. Since T_0 is preperfect, $uyu^{-1} \mid\!-^*\!-\!\mid vzv^{-1}$. If $u = 1$, then $y \mid\!-^*\!-\!\mid vzv^{-1}$, so, since y is cyclically minimal, $v = 1$ and $y \mid\!-^*\!-\!\mid z$. Continuing by induction on $|u|$, suppose $u = aw$, so that $awyw^{-1}\overline{a} \mid\!-^*\!-\!\mid vzv^{-1}$. If $|v|_a = 0$ then (using the property noted above) v is independent of a and there is some z_1 such that $z \;|\!-\!*\!-\!|\; az_1\overline{a}$, contradicting the assumption that z is cyclically minimal. Therefore, $v \;|\!-\!*\!-\!|\; at$ for some string t, and $wyw^{-1} \;|\!-\!*\!-\!|\; tzt^{-1}$. Both wyw^{-1} and tzt^{-1} are minimal (since they are substrings of minimal strings) and w is shorter than u, so $y \;|\!-\!*\!-\!|\; z$.■

The core of a string can be found in linear time by extending the procedure to find a minimal string congruent to a given string.

Lemma 8. For each free partially commutative group $G(\theta_0)$, there is a linear-time algorithm to find the core $\sigma(w)$ of a given string $w \in \Sigma^*$.

Proof. Given $w \in \Sigma^*$, the core of w can be found as follows. First, as in Theorem 5, form the tuple of strings $R(w) = (x_1, \ldots, x_N) = \Pi(x)$ where x is a minimal string congruent to w. With the strings $x_1, \ldots, x_N$ written on separate tapes, position a head at each end of each string. The sets $F(x)$ and $L(x)$ are evident from the letters under the heads, since letter a belongs to $F(x)$ if and only if it is the first letter of $\pi_i(x) = x_i$ for each i such that a belongs to A_i (and similarly for $L(x)$). If there is no letter $a \in F(x)$ such that $\overline{a}$ belongs to $L(x)$, then x is cyclically minimal and is the core of w. If there is such a letter a, then move both the heads on x_i inward one letter for each i such that $a \in A_i$; in effect, this replaces $\Pi(x)$ with $\Pi(y)$ where $x \;|\!-\!*\!-\!|\; ay\overline{a}$. This process can be repeated until a cyclically minimal string (in projected form) is obtained. ■

It is clear that strings are conjugate in $G(\theta_0)$ exactly when their cores are conjugate in $G(\theta_0)$. Moreover, as shown by the following lemma, cyclically minimal strings that are conjugate in $G(\theta_0)$ must be conjugate in the monoid $M(\theta)$.

Lemma 9. If x and y are cyclically minimal and $z^{-1}xz \equiv_G y$, then there exist u, v such that $xu \mid\!-^*-\!\mid uy$ and $vx \mid\!-^*-\!\mid yv$.

Proof. Since conjugacy is a symmetric relation in commutation monoids [12], the existence of either u or v implies the existence of the other.

If $x \equiv_G y$ then, since x and y are minimal, $x \mid\!-^*-\!\mid y$, so $u = v = 1$ will serve. Continuing by induction on $|z|$, suppose that $z^{-1}xz \equiv_G y$ with $z \neq 1$, where we may assume that z is a shortest such conjugator and, in particular, is minimal. Since y is minimal, $z^{-1}xz \mid\!-^*\!\rightarrow y$, and since y is cyclically minimal, it cannot be the case that $z^{-1}xz \mid\!-^*-\!\mid y$; hence there exist s and t such that $z^{-1}xz \mid\!-^*-\!\mid s \longrightarrow t \mid\!-^*\!\rightarrow y$. Since z and x are minimal but $z^{-1}xz \mid\!-^*-\!\mid s \longrightarrow t$, either
(i) $z \mid\!-^*-\!\mid az_1$ and $x \mid\!-^*-\!\mid ax_1$ for some $a \in \Sigma$; or
(ii) $z \mid\!-^*-\!\mid \overline{a}z_1$ and $x \mid\!-^*-\!\mid x_1a$ for some $a \in \Sigma$; or
(iii) $z \mid\!-^*-\!\mid az_1$ for some letter a such that x is independent of a.
(That is, the reduction either takes place at one end of x or across x.) The third case is excluded by the assumption that z is a shortest conjugator, and the second is symmetric with the first, so suppose the first case holds.

If $z \mid\!-^*-\!\mid az_1$ and $x \mid\!-^*-\!\mid ax_1$ then $z^{-1}xz \mid\!-^*-\!\mid z_1{}^{-1}\overline{a}ax_1az_1 \longrightarrow z_1{}^{-1}x_1az_1$, so, since $z^{-1}xz$ is congruent to y, $z_1{}^{-1}(x_1a)z_1 \mid\!-^*\!\rightarrow y$. The string x_1a is cyclically minimal because x is cyclically minimal and $x \mid\!-^*-\!\mid ax_1$, so (since z_1 is shorter than z) there exist u_1, v_1 such that $x_1au_1 \mid\!-^*-\!\mid u_1y$ and $v_1x_1a \mid\!-^*-\!\mid yv_1$. Let $u = au_1$ and $v = v_1x_1$; then $xu \mid\!-^*-\!\mid ax_1au_1 \mid\!-^*-\!\mid uy$ and $vx \mid\!-^*-\!\mid v_1x_1ax_1 \mid\!-^*-\!\mid yv$, as desired.■

Duboc has noted that if words over Σ_0^* (that is, containing only positive letters) are conjugate in $G(\theta_0)$, then there is a conjugator containing only positive letters [13, Prop. 3.3.8].

Combining Lemmas 8 and 9 with the fact that conjugacy in a commutation monoid can be tested in linear time, we arrive at the following conclusion.

Theorem 10. For each free partially commutative group $G(\theta_0)$, the Conjugacy Problem in $G(\theta_0)$ can be solved in linear time.

6. Other problems

The results presented in the previous sections suggest that properties and algorithms will generalize from free groups to the entire class of free partially commutative groups. To what extent does this "generalizing principle" hold? That is, what analogs of other properties of free groups hold for free partially commutative groups?

One example of such an analog concerns commuting elements. Elements in a free group that commute are powers (possibly negative powers) of a common element, and a similar result is true for free monoids. In the context of partial commutativity, the possibility that some parts of the strings are independent of each other must be taken into account. Based on the characterizations developed by Duboc [12] and Cori and Métivier [8] of commuting elements in commutation monoids, the following fact can be established.

Proposition 11. Minimal strings x and y commute in $G(\theta_0)$ if and only if $x \mid\!-^*\!-\!\mid u r_1^{j_1} \cdots r_m^{j_m} u^{-1}$ and

$y \mid\!-^*\!-\!\mid u r_1^{k_1} \cdots r_m^{k_m} u^{-1}$ where the strings r_i are pairwise

independent (and the exponents are integers).

Dehn's fundamental decision problems are the Word, Conjugacy and Isomorphism Problems. In this context the Isomorphism Problem takes the form: Given two partially commutative alphabets (Σ_0,θ_0) and (Σ_1,θ_1), are the groups $G(\theta_0)$ and $G(\theta_1)$ isomorphic? For a pair of free groups (that is, when both the independence relations θ_0 and θ_1 are empty), the groups are isomorphic exactly when the sets of generators Σ_0 and Σ_1 have the same size. Two commutation monoids are isomorphic exactly when their independence relations are "the same", that is, when there is a bijection between the alphabets that preserves the independence relations; this is because an isomorphism between the monoids must essentially map letters to letters. Another way to state this property is that the commutation monoids are isomorphic if and only if the (undirected) graphs of their independence relations are isomorphic. These observations suggest that the following property might hold for free partially commutative groups.

Conjecture. Two free partially commutative groups $G(\theta_0)$ and $G(\theta_1)$ are isomorphic if and only if the undirected graphs of their alphabets (Σ_0,θ_0) and (Σ_1,θ_1) are isomorphic.

REFERENCES

1. Aalbersberg, IJ. and Rozenberg, G., Theory of traces, Theor. Comp. Sci. 60 (1988), 1-82.
2. Aho, A. V., Hopcroft, J. E. and Ullman, J. D., The Design and Analysis of Computer Algorithms, Addison-Wesley (1974).
3. Book, R., Confluent and other types of Thue systems, J. ACM 29 (1982), 171-182.
4. Book, R., Thue systems as rewriting systems, J. Symbolic Computation 3 (1987), 39-68.
5. Book, R. and Liu, H.-N., Rewriting systems on a free partially commutative monoid, Info. Proc. Letters 26 (1987), 29-32.

6. Cartier, P. and Foata, D., Problemes Combinatoires de Commutation et Rearrangements, Lecture Notes in Mathematics 85, Springer-Verlag (1969).
7. Cochet, Y. and Nivat, M., Une généralization des ensembles de Dyck, Israel J. of Math. 9:3 (1971), 389-395.
8. Cori, R. and Métivier, Y., Recognizable subsets of some partially commutative monoids, Theor. Comp. Sci. 35 (1985), 179-189.
9. Cori, R. and Perrin, D., Automates et commutations partielles, RAIRO - Informat. Theor. 19 (1985), 21-32.
10. Diekert, V., On the Knuth-Bendix completion for concurrent processes, Automata, Languages, and Programming, Lecture Notes in Computer Science 267, Springer-Verlag, 1987, 42-53.
11. Duboc, C., Some properties of commutation in free partially commutative monoids, Info. Proc. Letters 20 (1985), 1-4.
12. Duboc, C., On some equations in free partially commutative monoids, Theor. Comp. Sci. 46 (1986), 159-174.
13. Duboc, C., Commutations dans les monoides libres, thesis, Univ. Rouen, 1986.
14. Flé, M. and Roucairol, G., Maximal serializability of iterated transactions, Theor. Comp. Sci. 38 (1985), 1-16.
15. Hopcroft, J. E. and Ullman, J. D., Introduction to Automata Theory, Languages and Computation, Addison-Wesley (1979).
16. Huet, G., Confluent reductions: Abstract properties and applications to term rewriting systems, J. ACM 27:4 (1980), 797-821.
17. Jantzen, M., Confluent String Rewriting, EATCS Monograph Series 14, Springer-Verlag (1988).
18. Jouannaud, J.-P. and Kirchner, H., Completion of a set of rules modulo a set of equations, SIAM J. Computing 15:4 (1986), 1155-1194.
19. Liu, H. N., Wrathall, C. and Zeger, K., Efficient solution of some problems in free partially commutative monoids, Information and Computation, to appear.
20. Lothaire, M., Combinatorics on Words, Addison-Wesley (1983).

21. Magnus, W., Karrass, A. and Solitar, D. , Combinatorial Group Theory, John Wiley/Interscience, 1966.
22. Mazurkiewicz, A., Traces, histories and graphs: instances of a process monoid, Proc. MFCS 84, Lecture Notes in Computer Science 176, Springer-Verlag (1984), 115-133.
23. Narendran, P. and McNaughton, R., The undecidability of the preperfectness of Thue systems, Theor. Comp. Sci. 31 (1984), 165-174.
24. Narendran, P. and Otto, F., Preperfectness is undecidable for Thue systems containing only length-reducing rules and a single commutation rule, Info. Proc. Letters, to appear.
25. Nivat, M. (with M. Benois), Congruences parfaites et quasi-parfaites. Séminaire Dubreil, 25th année, 1971-1972, 7-01-09.
26. Perrin, D., Words over a partially commutative alphabet, in A. Apostolico and Z. Galil (eds.), Combinatorial Algorithms on Words, NATO-ASI Series, Springer-Verlag, 1985, 329-340.
27. Sakarovitch, J., On regular trace languages, Theor. Comp. Sci. 52 (1987), 59-75.
28. Wrathall, C., The word problem for free partially commutative groups, J. Symbolic Computation 6 (1988), 99-104.

SOME RESULTS ON VLSI PARALLEL ALGORITHMS

Xu Meirui Zhao Dongyue Liu Xiaolin

Institute of Information Science
Beijing Computer Institute
Beijing, China

ABSTRACT. In this paper,some VLSI algorithms are presented.Based on the Knuth's algorithms,we develop a parallel algorithms on VLSI mesh of trees organization to solve the optimal binary search tree problem. Its time complexity is $O(n \log n)$ and the area is $O(n^2 \log^2 n)$.We also propose some VLSI parallel algorithms for some feature extraction operations such as linear edge enhancement and nonlinear edge enhancement.

1. Introduction. It is well known that there is only one general- purpose processor in the traditional (Von Neumann Scheme) computer on which people design different sequential algorithms to solve different problems. But now,the case has been changed. The advent of VLSI has led to the development of high performance, special-purpose hardware to meet many specific application requirements. With VLSI technology, people could make full use of the inherent parallelisms in the problems, then find corresponding processor array organization which is very suitable to implement these parallelisms. Since 1978, a number of organizations have been devloped, such as systolic array, wavefront array, mesh organization, tree organization,mesh of trees, hypercube, CCC network, shuffle- exchange organization, etc . In the meantime, many efficient parallel algorithms on these organizations have been proposed.

In this paper, a parallel algorithm to solve the optimal binary search tree is presented. The algorithm is supported by the mesh of trees organization. In section 2, a basic description about the mesh of trees organization and the optimal binary search tree problem (OBST) is given. Section 3 is devoted to the VLSI OBST algroithm. Two algorithms for edge enhancement are proposed in section 4.

2. The mesh of trees organiztation and the optimal binary search tree problem. The mesh of trees of side n, where n is a power of 2, is defined as follows (see[7] [10])

1) There are n^2 nodes arranged in a square grid, but without any connecting edges. The nodes in position (i,j) is called P_{ij} ,$1 \leq i,j \leq n$.

2) For each row of nodes, build a complete binary tree with them as leaves,adding the interior nodes,which are not among the original n^2 nodes.The i-th row tree is notated $T_{ri}, 1 \leq i \leq n$.

3) Do the same work for each column,then we get column tree $T_{ci}, 1 \leq i \leq n$.

Fig.1 shows the mesh of trees of side 4.If we substitute processors for nodes and buses for edges in the mesh of trees, then we get the mesh of trees organization (MT,for short).We also very often regard the root of T_{ri} and the root of T_{ci} as the same processor,and call it the i-th controller (C_i).

We enumerate some of the operations for which MT are especially suitable.These operations include the following.

1) **Broadcasting.** A data at C_i can be sent to all leaf processors of T_{ri} (or T_{ci}) along the i-th row tree (or the i-th column tree) in time O(log n);

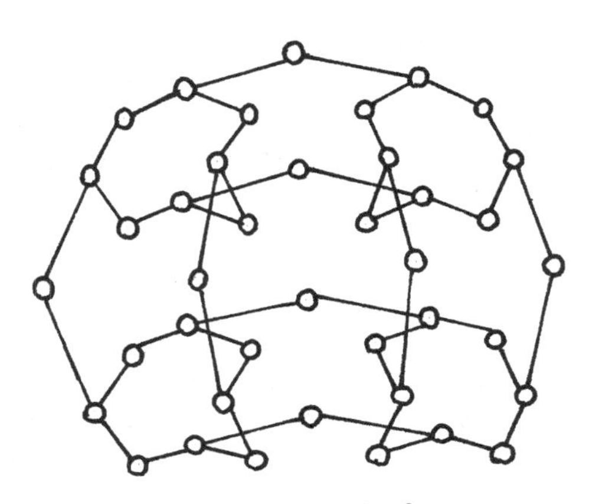

Fig.1 The mesh of tree

2) **Send-up.** A leaf processor P_{ij} sends a data to C_i via T_{ri}, or to C_j via T_{cj}. Its time is O(log n);

3) **Send-down** (Selected broadcasting). A data at C_i can be sent to some leaf processors,say the s-th to the t-th , of T_{ri} (or T_{ci}, ($1 \leq s,t \leq n$).The time is also O(log n).

4) **Census function.** A census function is a commutative,associative operation on n values,each at one of the leaves of a tree (T_r or T_c),such as addition,multiplication,minimum,maximum,logical 'or' and logical 'and',etc. They can be performed in O(log n) time provided that each operation takes O(1) time.

Now we turn to the optimal binary search tree problem.Suppose that we are given a set $S=\{a_1,a_2,\dots,a_n\}$,that is, a subset of some large universal set U which is linearly ordered by a relation "≤".In addition,we assume $a_1 < a_2 < \dots < a_n$. Th e set S can be represented by a binary search tree (BST), in which each vertex v is labeled by an element $l(v) \varepsilon S$,such that for each vertex u in the left subtree of v,$l(u) = l(v)$,for each vertex w in the right subtree of v,$l(w) = l(v)$,and for each element $a \varepsilon S$,there is exactly one vertex v such that $l(v)=a$.Let P_i be the probability of searching ai qi be the probability of searching $a \varepsilon U$, $a_i < a < a_{i+1}$ for some a_i and q_0,q_n be the probability of searching $a, a < a_1$ and $a_n < a$, respectively. We also add n+1 fictitious leaves to the BST to reflect the elements in U-S . They are labeled 0,1,...,n [1]. The cost of a BST is defined as

$$\mathbf{cost} = \Sigma_i \ (p_i * (\mathbf{Depth}(a_i)+1)) + \Sigma_j \ (q_j * \mathbf{Depth}(j))$$

where $1 \leq i \leq n$, $0 \leq j \leq n$. The optimal binary search tree of S is a minimum-cost BST.

How to construct the OBST when the set S and P_i $(1 = i = n)$,q_i $(0 = i = n)$ are given ? That is so called OBST problem.For this problem, Gilbert and Moor gave a well known algorithm in 1959. [3] The time complexity of the algorithm is O (n^3) if we assume that one addition (or comparison) takes O(1) time.In this paper,for convenience,we always make the same assumption.Since otherwise we need only add a factor $\log^k n$ to the time complexity . Now ,let T_{ij} be a minimum cost tree for the subset $S_{ij} = \{a_{i+1}, a_{i+2}, \ldots, a_j\}$, and C_{ij} be the cost of T_{ij},R_{ij} be the number of the root of T_{ij}. The algorithm cousists of two procedures. The first one computes R_{ij} and C_{ij} ,for $0 \leq i \; j \leq n$ while the second constructs OBST for S (i.e T_{on}) recursively.The first procedure is as below.

```
        PROCEDURE
          begin
1           for i := 1 to n do
             begin
2              wii := qi ;
3              rii := i   ;
4              cii := 0
             end ;
5           for l := 1 to n do
6             for i := 0 to n-l do
               begin
7                j := i+l ;
8                wij := wi,j-1 + pj + qj ;
9                let m be a value of k, i < K <= j , for which ci,k-1 + ckj is minimum ;
10               cij :=   wij + ci,m-1 + cmj
11               rij := m
               end
        end
```

The second procedure Buildtree is as below.

```
    PROCEDURE  Buildtree ( i ,j )
      begin
        create vertex rij, the root of Tij ;
        if i < rij-1 then
            make Buildtree (i,rij-1), the left subtree of rij;
        if rij<j then
            make Buildtree (rij,j), the right subtree of rij
      end
```

An improvement of this algorithm was made by Knuth in 1971 [5]. Knuth proved that the algorithm still works if the inequality $i < k \leq j$ in line 9 is substituted by $\max(i+1, r_{i,j-1}) \leq k \leq r_{i+1,j}$.The time of the new algorithm is $O(n^2)$.

Based on Knuth's algorithmsr,a VLSI algorithm on mesh of trees orgaization for the problem OBST is developed.Its time complexity is $O(n \log n)$.

3).VLSI OBST Algorithm. For our OBST algorithm, we use the mesh of trees organization of side $n+1$,where $n+1$ is a power of 2, and suppose that the root of T_{ri} and the root of T_{ci} are the same (i.e C_i).For each leaf processor P_{ij} ($0 \leq i,j \leq N$), there are seven registers in it , they are P, Q, W, C, C',MARK and R.During the procedure, the main work of P_{ij} is to compute wij,cij and rij. Of course, P_{ij} will be used to pass data sometimes.

There are some observations which can be made by analyzing the Knuth's algorithm and the mesh of trees organization.

1. The functions from line 6 to line 11 in the first procedure can be computed in parallel for each l ($1 \leq l \leq n$). That is, as shown in Fig.2, in our OBST algorithm the sequence of parallel computation is the following :

1) $P_{00}, P_{11}, P_{22}, \ldots, P_{nn}$ (in parallel);
2) $P_{01}, P_{12}, P_{23}, \ldots, P_{n-1,n}$ (in parallel);

......

n+1) P_{0n}

2. In the 1-th parallel step,for each P_{ij} ($j=i+1$) the most important work is to compute

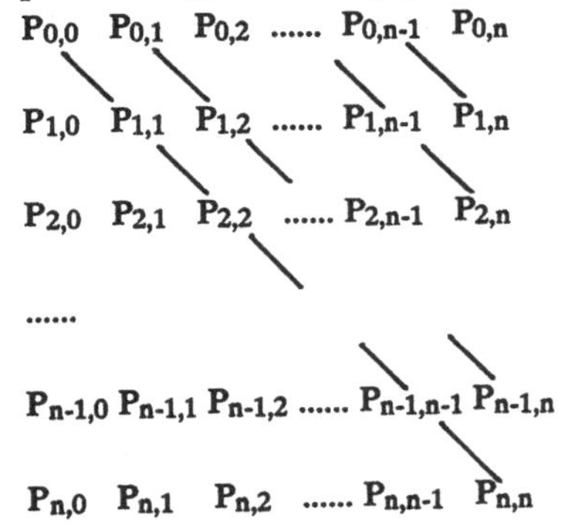

Fig.2 Parallel Computation Schema

$\min_k (c_{i,k-1} + c_{k,j})$, where $\max\{i+1, r_{i,j-1}\} \leq k \leq r_{i+1,j}$. It is clear that the value $c_{i,k-1}$ in $P_{i,k-1}$ must be transfered to P_{kj} to perform the addition $c_{i,k-1} + c_{k,j}$ for all k in above interval. To do this, there are two possible paths. One is so called CR path, the other is RC path (Fig.3) . For example , the CR path is as follows:

$P_{i,k-1}$----------C_{k-1}----------$P_{k,k-1}$-----------C_k--------$P_{k,j}$

$T_{c,k-1}$ $T_{c,k-1}$ $T_{r,k}$ $T_{r,k}$

(send up) (send down) (send up) (send down)

In general case, we expect to use CR paths to send the c's of all ralative rows to their cor-

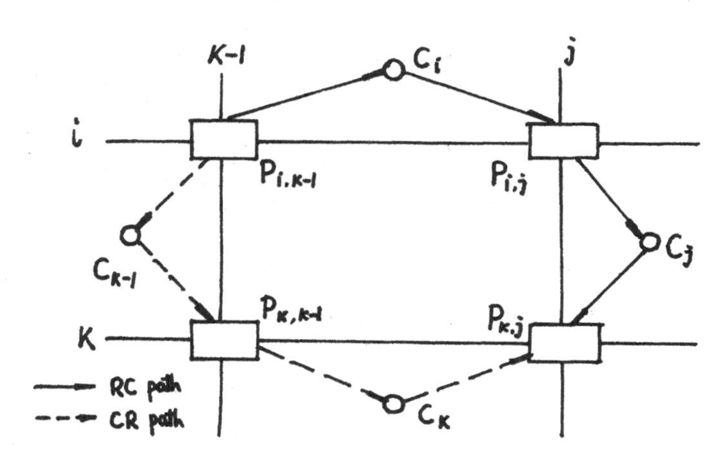

Fig.3 RC path and CR path

responding position of various columns with time O(log n). The problem is that the data overlap might happen when they are traveling along their own CR paths. In fact, we can find the data overlap from the following inequalities:

$\max\{i+1, r_{i,j-1}\} \leq k \leq r_{i+1,j}$ for row i,

$\max\{i+2, r_{i+1,j}\} \leq k \leq r_{i+2,j+1}$ for row i+1.

Fortunately, we can find another way to rearrange the transfer paths to avoid these data overlaps. By looking at above two inequalities, we know that there is only one data in each row which will encounter one other data in the next row during their travels along the CR paths(See Fig.4)

To solve this problem, one way is to force one of the overlap data to be delayed for O(1) time. But there is another way which will be more regular and elegant. That is, let the number c in last processor (i.e $P_{i,k-1}$, $k=r_{i+1,j}$) in each row transfer to the destination by the RC path:

$P_{i,k-1}$--------C_i--------$P_{i,j}$--------C_j--------$P_{k,j}$

T_{ri} T_{ri} T_{cj} T_{cj}

where $k=r_{i+1,j}$. It will be obvious that there is no data conflict on any column trees in the same parallel step, and neither on any row trees, even though in the case that $r_{i,j-1}=r_{i+1,j}$, since there is only one data in each row which will go along the RC path. (Fig.5)

Now, The parallel algorithm for constructing optimal binary search tree is given as below. Without loss of generality, We suppose that each P_{ij} and C_i know its own number in the array. The first procedure computes c_{ij} and r_{ij}, and the second outputs T_{on}.

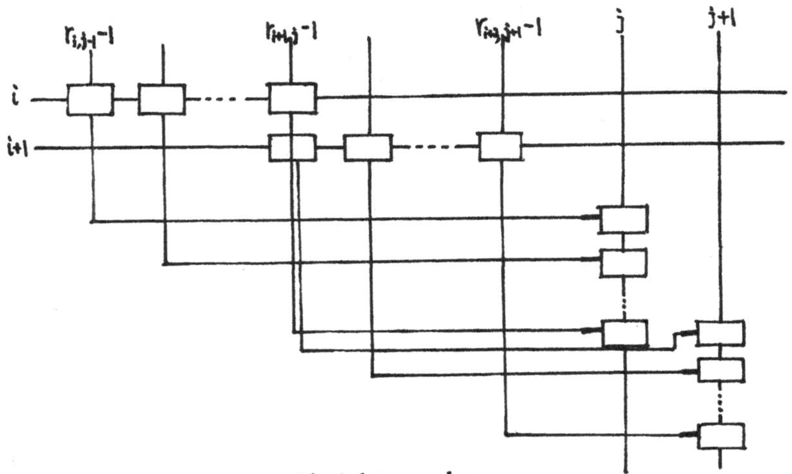

Fig.4 data overlap

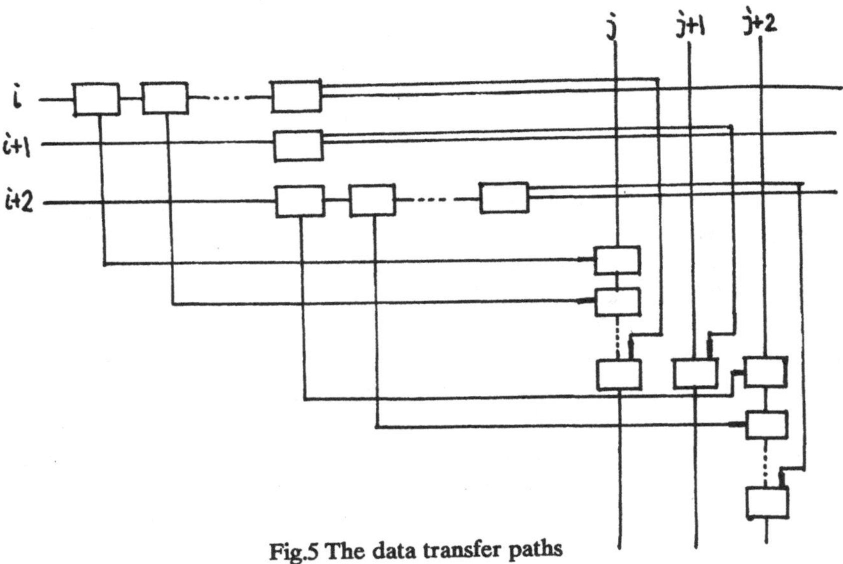

Fig.5 The data transfer paths

ALGORITHM OBST

0 Initialize by reading the p_j, q_j into C_j ($1 \leq j \leq n$) and q_0 into C_0, Broadcasting p's and q's using $T_{c's}$.AS a result, each P_{ij} holds p_j and q_j, and P_{0j} holds q_0 ($0 \leq i \leq n$, $1 \leq j \leq n$).

1 For all Pii ($0 \leq i \leq n$) DO (in parallel)

```
    BEGIN
      wii: =qi; cii: =0; rii: =i
    END
2    For l: =1 to n Do
    BEGIN
3        FOR all Pij (0 ≤ i,j ≤ n) DO (in parallel)
          IF i+l-1=j THEN
            send wij to Pi,j+1 using the path
            Pij---------Ci---------Pi,j+1   ;
                Tri       Tri
4        FOR all Pij 3(0≤i,j≤n) DO (in parallel)
          IF i+l=j THEN
            wij: =wi,j-1 + qj + pj   ;
5        FOR all Pij (0 ≤ i,j ≤ n) DO (in parallel)
          BEGIN
          IF i+l-1=j  THEN
            send-up (rij) to Ci, using Tri   ;
            IF (i+l-1=j) AND (i>0) THEN
              send-up(rij) to Ci-1, using the path
          Pij---------Cj---------Pi-1,j---------Ci-1
              Tcj       Tcj            Tr,i-1
        /* therefore ci holds ri,j-1 and ri+1,j */
          END    ;
6    FOR all Ci (0  ≤ i ≤ n-l) DO (in parallel)
        BEGIN
        send-down Signal 0 to set 01 to MARK of
        Pi,k-1 , where k=ri+1,j   ;
        send-down Signal 1 to set 10 to MARK of
        Pi,k-1, where max {i+1, ri,j-1} ≤ k ≤ ri+1,j
        END    ;
7    FOR all Pik (0 ≤ i,k ≤ n) DO (in parallel)
      BEGIN
        IF MARK of Pik is 01 THEN
         dismiss MARK 01 and send cik to c' of
         Pk+1,j, using RC path, and set 01 to MARK of
         Pk+1,j ;
        IF MARK of Pik is 10 THEN
          dismiss MARK 10, and send cik to c' of Pk+1,j
          using CR path,and Set 01 to MARK of Pk+1,j
      END;
 8  FOR all Pk,j (0 ≤ k,j ≤ n) DO (in parallel)
```

IF MARK of $P_{k,j}$ is 01 THEN
Compute $c := c + c'$;
/* befor computing, $c = c_{kj}$, $c' = c_{i,k-1}$ */
9 **FOR** all C_j $(1 \leq j \leq n)$ **DO** (in parallel)
BEGIN
Compute m_j and s_j, using $T_{c,j}$,Where
$s_j := \min \{ c \mid c$ is from the leaf processor P,
MARK of which is 01 $\}$
m_j is the value t such that $s_j = C_{i,t-1} + C_{t,j}$;
dismiss MARK 01 :
send-down s_j and m_j to $P_{i,j}$, where $i=j-1$
END;
10 **FOR** all P_{ij} $(0 \leq i,j \leq n)$ **DO** (in parallel)
IF $i+1=j$ **THEN**
BEGIN
$c_{ij} := w_{ij} + s_j$;
$r_{ij} := m_j$
END
END

Then second procedure (see below) will output the optimal binary search tree (i,e T_{on}) in the following way. First,it will output the root of T_{on},then output all nodes of depth 1 inparallel,then all nodes of depth 2,and so on.Since for every two Subtree $T_{i,j}$ and $T_{k,l}$ of T_{on},if their roots are at the same level, then $\{i+1,...,j\} \cap \{k+1,...,l\} = 0$, therefore,there is no data overlap in second procedure.

BEGIN
set 01 to MARK of P_{on};
REPEAT n times **DO**
FOR all $P_{i,j}$ $(0 \leq i,j \leq n)$ **DO** (in parallel)
IF MARK of $P_{i,j}$ is 01 **THEN**
BEGIN
dismiss MARK 01;
Send-up $(i,j,r_{i,j})$ to C_i to output,using $T_{r,i}$;
IF i $r_{i,j}$ - 1 **THEN**
sent signal to set 01 to MARK of $P_{i,k}$
$(k=r_{i,j}-1)$,using the path
$p_{i,j}$--------C_i--------$P_{i,k-1}$;
$T_{r,i}$ $T_{r,i}$
IF $r_{i,j}$ j **THEN**
send signal to set 01 to MARK of $P_{k,j}$
$(k = r_{i,j})$,using the path

$P_{i,j}$--------C_j--------$P_{k,j}$

$T_{c,j}$ $T_{c,j}$

END

END

Theorem. The above parallel algorithm is correct.The time complexity of the alorithm is $O(n \log n)$,and the area is $O(n^2\log^2 n)$.

Proof: The correctness of the algorithm is from above analysis. Then we discuss the time complexily in detail.As noted above, we assume that the addition and comparison both take O(1) time,and every processor needs O(1) area.In the first algorithm the costliest part is from line 2 to line 10. The loop at line 2 is executed n times. Each step in the loop of line 3--10 requires at most O(log n) time. so the total time for first procedure is O(log n).It is easy to see that the second procedure requires only O(n log n) time.

A mesh of trees of side n+1 requires $O((n+1)^2 \log^2(n+1))$ area [7],[10],which is equal to $O(n^2 \log^2 n)$.

4. VLSI algorithms for feature extraction. Now,we turn to some feature extraction algorithms.Feature extraction is an important procedure for pattern recognition and image analysis,and includes some operations such as edge detetion,histogram feature extraaction,transform coeficient feature extraction and texture feature extraction,etc.In recent years,many of these operations have been implemented by some VLSI special architectures.Systolic arrays for edge detection have been presented [9];non--systolic architecture for edge detection proposed by Offen [9].Huang and SU designed a matrix arithmetic network to implement the Foley-sammon feature select algorithm [4]. Modgera described the method of VLSI implementation of optimal feature selection [8].The systolic arrays for DFT and FFT have been proposed in [2].

The edges of image are important feature of an image.In general,edge detection includes edge enhancement and threshold detection.Now,we shall propose two edge enhancement VLSI algorithms,which can be serves as a part of an image processing system.

One of the simple techniques of edge enhancement is discrete differenciation of image.Suppose that the original image is a n x m pixel matrix $f=\{f(i,j) \mid 1 \leq i \leq n, 1 \leq j \leq m\}$.By selecting proper mask

$$A: \quad \begin{matrix} a(1,1) & a(1,2) & a(1,3) \\ a(2,1) & a(2,2) & a(2,3) \\ a(3,1) & a(3,2) & a(3,3) \end{matrix}$$

we obtain a output image G according to the relation

$$G(i,j) = E_l\, E_k \quad a(l,k)*F(i+l-2,j+k-2),$$

where $1 \leq l, k \leq 3$ and $F(i,j)=0$ if $i=0$ or $j=0$. G is called the impage obtained by linear edge enhancement.

The processor array and data I/O pattern for imag linear edge enhancement (ILEE)are given in Fig.6,and the inner construction of each processor is given in Fig.7.In each beat,every processor does the same work according to the following algorithm:

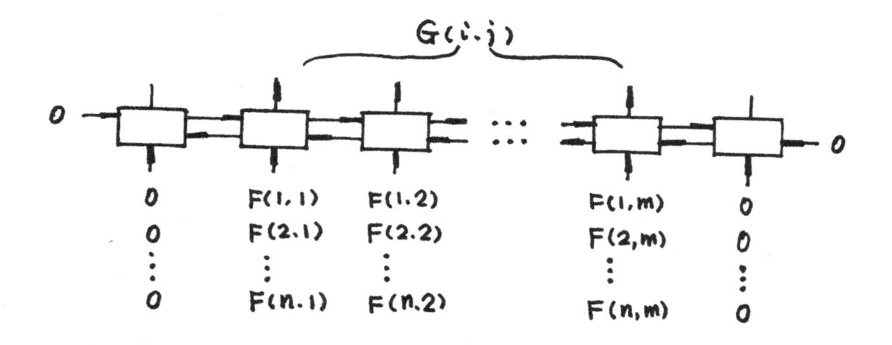

Fig.6 The processor array for ILEE

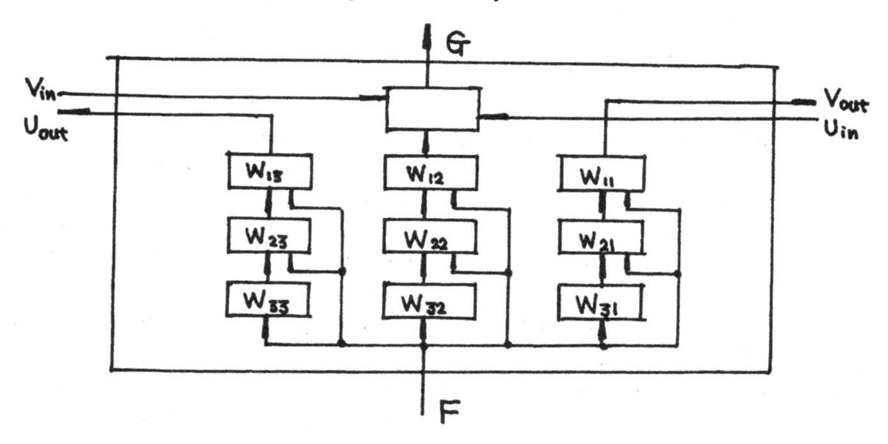

Fig.7 The inner construction of processor

ALGORITHM ILEE

beat begin

$G := W_{12} + U_{in} + V_{in}$;

for all W_{ij} $(1 \leq i \leq 2, 1 \leq j \leq 3)$ **do** (in parallel)

$W_{ij} := W_{i+1,j} + a(4-i,j)*F$;

for all W_{3j} $(1 \leq j \leq 3)$ **do** (in parallel)

$W_{3j} := a(1,j)*F$

beat end

In the algorithm ILEE,three sentances are implemented in parallel.The time complexity of the algorithm ILEE is O(n),and the area is O(m) since all a's are coustant and the value of function F is bounded by some constant.

If the following relation is used to transform our image,then the resulting image is called the image obtained by nonlinear edge enhancement (INLEE) :

$$G(i,j) = SQR((F(i,j)-F(i+1,j+1))^2 + (F(i,j+1)-F(i+1,j))^2).$$

The processor array and the data I/O pattern for INLEE are shown in Fig.8 a) and the inner con-

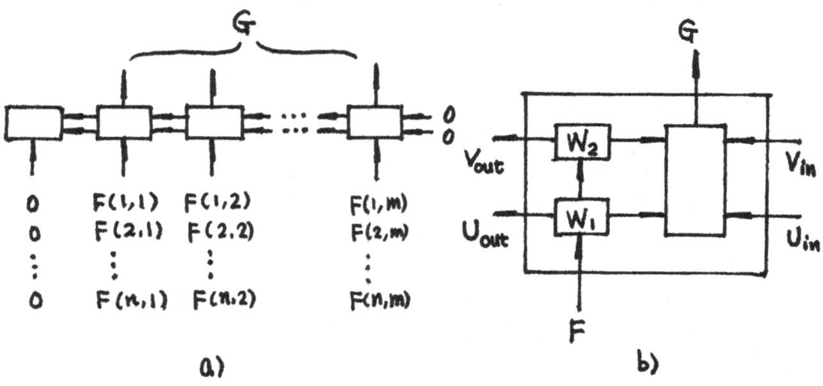

Fig.8 The structure for INLEE

struction of each processor is depticted in Fig.8 b) . In each beat, every processor does the same work according to the following algorithm:

ALGORITHM INLEE

beat begin

$G:=SQR((W_1 - U_{in})^2 + (V_{in} - W_2)^2);$

$W2:=W1$;

$W1:=F$

beat end

also,the time of the algorithm INLEE is O(n),and the area is O(m),as all value F is bounded by some constant.

At the end,we want to make a few observations on the use of previous two algorithms and array structures.

1. It should be clear that if there is only one array structure(say,ILEE or INLEE) on a chip,this design is inefficient. A suggested way to construct a chip system is given below.Since during the process of pattern recognition or image processing,an image will under-

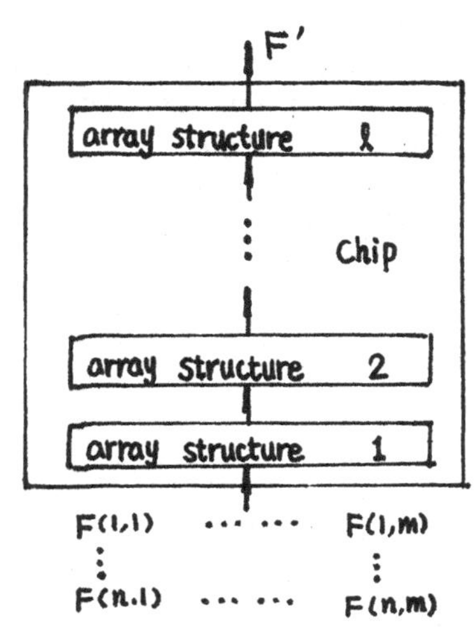

Fig.9

go a sequence of transformations,so we may find some array structure for each transformation,then put all such array structrues on a chip system one after another with the original image as the input and the final resulting image of these transformations as the output.(Fig.9)

2. The number m may be sufficient large,and the number of input tip for a chip is limited;therefore we must patition the whole array structure system into several subsystems in such a way that each subsystem deal only with a sub-image of size n x k (k x t = m for some t). We need only design one chip for each subsystem with the data flow arranged in a proper way.

REFERENCE

[1] Aho,A.V.,Hopcroft,J.E. and Ullman, J.D., The Design and Analysis of Computer Algorithms, Reading MA: addison-Wesly, 1974.

[2] Gertner, I. and Shamash,M. VLSI Architecture for Multidimensional Fourier Transform Processing, IEEE Trans. On Computers, Vol. C-36,no.11,1987,1265-1274.

[3] Gilbert,E.N. and Moore,E.F. , Variable lenghth encodings,Bell System Technical J. 38:4,1959, 933-968.

[4] Huang,K. and Su,S.P. , VLSI Architecture for Feature Extraction and Pattern Recognition , Computer Vision and Image Processing , 24,1983,215-218.

[5] Knuth, D.E. , Optimum Binary Search Tress , Acta Informatica 1,1971,14-25.

[6] Kung, H.T. and Card, L.P. , One Dimension Systolic Arrays for multidimensional Convolution and Resampling, in VLSI for Pattern Recognition and Image Processing,K.S. Fu eds., Springer Verlag,1984.

[7] Leighton,F.T. ,New Lower Bound Techniques for VLSI, proc.22nd Annual IEEE Symp. FOCS, 1981,1-12.

[8] Modgera, S.D. , Toward a fundamental theory of optimal feature selection: Part II --- implement and computational complexity, IEEE Trans, on Pattern Analysis and Machine Intelligence, Vol.PAMI-9, no.1 ,1987.

[9] Offen,R.J., VLSI image Processing , William Collins Sons & CO.,London,1985, 122-123.

[10] Ullman,J.D. , Computational Aspects of VLSI , Comput. Sci. Press,Inc. Rockville Maryland,1984.

ON OPTIMAL ARRANGEMENT OF 12 POINTS

Yao Tianxing

Department of Mathematics
Nanjing University
Nanjing, PRC

Abstract

We arrange n distinct real numbers according to their values, and let r(n) be the least number such that at most r(n) comparisons are needed in any cases. It is well-known that the lower bound of r(n) is $m(n)=1+[\log_2 n!]$, for $n>2$. Steinhaus had conjectured: $r(n)=m(n)$ when $n>2$. Ford and Johnson gave a program of comparisons which consists of FJ(n) comparisons, and they conjectured: r(n)=FJ(n). For $n<12$, it has already been verified that m(n)=FJ(n), while m(12)= 29, FJ(12)=30. In this article we proved r(12)=30 by means of manual computations.

We arrange n distinct real numbers according to their values by pairwise comparisons. Let r=r(n) be the least positive integer such that there exists a method of comparisons, in which at most r comparisons are necessary for the arrangement of these n numbers.

Let $n=2^{k-1}+q$, $0\leq q<2^{k-1}$. Steinhaus (1950) has proved

$$r(n)\leq S(n) \equiv 1+nk-2^k.$$

Ford and Johnson (1959) have shown that the lower bound of r(n) is

$$m(n) \equiv 1+[\log_2 n!]$$

for $n>2$, and have obtained a method of comparisons. Let FJ(n) be the number of these comparisons. Then $FJ(n)<S(n)$, when $n>4$. Ford and Johnson had conjectured that

$$r(n) = FJ(n).$$

But Steinhaus conjectured that for $n>2$,

$$r(n) = m(n).$$

Here we give the following table of m(n), FJ(n) and S(n) ($n\leq 26$).

n	1	2	3	4	5	6	7	8	9	10	11	12	13
m(n)	0	1	3	5	7	10	13	16	19	22	26	29	33
FJ(n)	0	1	3	5	7	10	13	16	19	22	26	30	34
S(n)	0	1	3	5	8	11	14	17	21	25	29	33	37

n	14	15	16	17	18	19	20	21	22	23	24	25	26
m(n)	37	41	45	49	53	57	62	66	70	75	80	84	89
FJ(n)	38	42	46	50	54	58	62	66	71	76	81	86	91
S(n)	41	45	49	54	59	64	69	74	79	84	89	94	99

From this table we may find $r(n)=FJ(n)=m(n)$ when $1\leqslant n\leqslant 11$, or $n=20, 21$. But $r(n)$ are unknown for $12\leqslant n\leqslant 19$, and $n\geqslant 22$.

In this paper we shall show $r(12)=30$. First, let us introduce some notions.

We call a digraph without cycles and multiple arcs to be a <u>partial graph</u>. Let $G=G(V(G),A(G))$ be a partial graph. We add some arcs into G so that G becomes a transitive graph. If we can obtain w transitive graphs with different order type of vertices, then we say that the <u>potential</u> of G is $w=w(G)$. A transitive graph obtained by adjoining arcs into G is called a transitive graph of G. Let E and T be an empty graph and a transitive graph of order n respectively. Obviously $w(E)=n!$, $w(T)=1$.

Let G be a partial graph and $u,v\in V(G)$. We say that u,v constitute an <u>orderless pair</u>, if both $\overrightarrow{uv}$ and $\overrightarrow{vu}\,\bar{\in}\,A(G)$. In this case, adding arc $\overrightarrow{uv}$ into G, the partial graph thus obtained according to transitive relation is denoted by $G+\overrightarrow{uv}$. We say that $G+\overrightarrow{uv}$ is the partial graph obtained from G by the <u>order-adjoining</u> of $\overrightarrow{uv}$. Obviously

$$w(G) = w(G+\overrightarrow{uv}) + w(G+\overrightarrow{vu}).$$

We say that $\overrightarrow{uv}$ is a <u>major arc</u> of G if $w(G+\overrightarrow{uv})\geqslant w(G+\overrightarrow{vu})$. Let

$$p(G,\overrightarrow{uv}) = 2w(G+\overrightarrow{uv})/w(G),$$

$$p(G,uv) = \max\left\{p(G,\overrightarrow{uv}),\ p(G,\overrightarrow{vu})\right\}.$$

We say that $p(G,\overrightarrow{uv})$ is the <u>partial ratio</u> of the adjoining arc $\overrightarrow{uv}$ into G, and $p(G,uv)$ is the <u>major partial ratio</u> of the orderless pair u,v in G. Obviously $p(G,uv)\geqslant 1$. And if there exists an automorphism σ

of G such that $\sigma(u)=v$, then $p(G,uv)=1$.

Let E be an empty graph of order n. After t times of successive order-adjoining of arcs between the orderless pairs of vertices, E becomes a transitive graph. The corresponding partial graph are G_1, $G_2,\ldots,G_t$, where G_t is a transitive graph. Let

$$p_i = 2w(G_i)/w(G_{i-1}),\ i=1,2,\ldots,t.$$

We say that $p_1,p_2,\ldots,p_t$ is a partial ratio sequence of order n. By the previous definitions we have

$$p_1p_2\cdots p_t = \frac{2w(G_1)}{w(E)}\cdot\frac{2w(G_2)}{w(G_1)}\cdots\frac{2w(G_t)}{w(G_{t-1})} = \frac{2^t}{n!}.$$

When $p_i\geq 1$, $i=1,2,\ldots,t$, we say that $p_1,p_2,\ldots,p_t$ is a major partial ratio sequence of order n. Thus we have

Theorem 1 The major partial ratio sequence of order n, $p_1,p_2,\ldots,p_t$, satisfies

$$p_1p_2\cdots p_t = 2^t/n!.$$

When n=12, the minimal integer t satisfying $2^t/n!>1$ is 29. We have $2^{29}/12!=1.1208\ldots<1.121$. Assume $r(n)=29$. Then there exists a major partial ratio sequence of order 12, $p_1,p_2,\ldots,p_{29}$, such that $p_1p_2\cdots p_{29}<1.121$. By the exhaustion method we have proved that there doesn't exist any major partial ratio sequence for n=12, t=29. To start with the empty graph E, we have discussed it according to the ascending order of the number of vertices of the maximum components and have checked up all cases by means of manual computations.

We usually have to compute the potential of a partial graph and the major partial ratio of orderless pairs of the vertices. The following theorems and formulas considerably simplify these computations, so that such manual computations become possible.

We say that v is a maximal member of G, if there doesn't exist such a vertex u that $\overrightarrow{uv}\in A(G)$. And we say that v is a greatest member of G, if $\overrightarrow{vu}\in A(G)$ for all $u\in V(G)$, $u\neq v$. Similarly we may define the mimimal and least member of G.

Theorem 2 (Maximal and minimal method) Let $v_1,v_2,\ldots,v_k$ be all maximal (or minimal) members of partial graph G. Then

$$w(G) = \sum_{i=1}^{k} w(G-v_i).$$

<u>Proof</u> Every transitive graph of G obtained by order-adjoining has a greatest member $v_i (1 \leqslant i \leqslant k)$. And the transitive graphs with a greatest member v_i or v_j are different for $1 \leqslant i \leqslant j \leqslant k$. So the theorem holds.

Theorem 2 gives a method for computing the potential of G. But there are more convenient methods for some special partial graphs (see the following Theorems 3 and 4).

Let F be an induced subgraph of G and $u \in V(G)$, $u \bar{\in} V(F)$. We say that u <u>dominates</u> over F (or u <u>is dominated</u> by F) if $\overrightarrow{uv} \in A(G)$ (or $\overrightarrow{vu} \in A(G)$) for all $v \in F$. We say that F is an <u>ordinal subgraph</u> of G if every vertex in G (but no in F) dominates over F or is dominated by F. We say that F is a <u>semi-ordinal subgraph</u> of G if for any $u \in V(G-F)$, there exists $v \in F$ such that $\overrightarrow{uv} \in A(G)$ ($\overrightarrow{vu} \in A(G)$), then u dominates over F (or is dominated by F).

Obviously we have the following Theorems 3 and 4.

<u>Theorem 3</u> (Method of ordinal subgraph) Let F be an ordinal subgraph of the partial graph G. We have

$$w(G) = w(F)w(G-F).$$

<u>Theorem 4</u> (Method of components) Let F and H be two disjoint subgraphs of a partial graph G and $G=F \cup H$, $|V(F)|=m$, $|V(H)|=n$. We have

$$w(G) = w(F)w(H)\binom{m+n}{m}.$$

We now give several methods for computing or estimating partial ratios.

<u>Theorem 5</u> (Method of semi-ordinal subgraph) Let F be a semi-ordinal subgraph of a partial graph G and u,v be an orderless pair of F. We have

$$p(G,\overrightarrow{uv}) = p(F,\overrightarrow{uv}).$$

<u>Proof</u> We add some arcs into F such that F becomes a transitive

graph and G becomes G'. Then

$$w(G) = w(F)w(G'),$$

$$w(G+\overrightarrow{uv}) = w(F+\overrightarrow{uv})w(G').$$

So

$$p(G,\overrightarrow{uv}) = 2w(G,\overrightarrow{uv})/w(G) = 2w(F+\overrightarrow{uv})/w(F) = p(F,\overrightarrow{uv}).$$

The conclusion is obtained.

<u>Theorem 6</u> (Method of the greatest member of components) Let F and H be two connective components of the partial graph G and $V(F)=m$, $V(H)=n$. And let u,v be the greatest members of F and H respectively. Then we have

$$p(G,\overrightarrow{uv}) = \frac{2m}{m+n}.$$

<u>Proof</u> By Theorem 4, $w(G')=w(F)w(H)\binom{m+n}{m}$, where $G'=F\cup H$. And

$$w(G'+\overrightarrow{uv}) = w(G'-u) = w(F-u)w(H)\binom{m+n-1}{n}$$

$$= w(F)w(H)\binom{m+n-1}{n}.$$

Obviously G' is a semi-ordinal subgraph of G. By Theorem 5 we have

$$p(G,\overrightarrow{uv}) = p(G',\overrightarrow{uv}) = 2w(G'+\overrightarrow{uv})/w(G')$$

$$= 2\binom{m+n-1}{n}\Big/\binom{m+n}{m} = \frac{2m}{m+n}.$$

<u>Theorem 7</u> Let u,v and u,w be the orderless pairs of a partial graph G and $\overrightarrow{vw}\in G$. We have

1) $p(G,\overrightarrow{uv})\geqslant 1 \Rightarrow p(G,uw)>p(G,uv)$;

2) $p(G,\overrightarrow{wu})\geqslant 1 \Rightarrow p(G,uv)>p(G,uw)$.

<u>Proof</u> 1) Obviously $G+\overrightarrow{uv} = (G+\overrightarrow{uw})+\overrightarrow{uv}$. Thus

$$w(G+\overrightarrow{uv}) = w((G+\overrightarrow{uw})+\overrightarrow{uv}) < w(G+\overrightarrow{uw}),$$

$$p(G,uw)\geqslant\frac{2w(G+\overrightarrow{uw})}{w(G)}\geqslant\frac{2w(G+\overrightarrow{uv})}{w(G)} = p(G,uv).$$

2) Similar to 1).

In order to prove $r(12)=30$ by the method of exhaustion, we have to deal with a huge amount of computations. The process is understood.

References

[1] Steinhaus, H., Mathematical Snapshots, Oxford University Press, (1950).

[2] Ford, L.R. and Johnson, S.M. A tournaments problem, Amer. Math. Monthly, (1959), 66, 387-389.

[3] Steinhaus, H., Some remarks about tournaments, Golden Jubilee Commemoration Vol. (1958/1959), Calcutta Math. Soc. 323-327.

SOME DISCUSSIONS ON VEHICLE ROUTING PROBLEMS

Yu Wenci*

Department of Statistics &
Operations Research
Fudan University
Shanghai,China

§1. Introduction

There have been many research works on vehicle routing problems, as summarized in survey paper [1]. Vehicle routing problems can be briefly described as follows. Suppose that there are m vehicles, one depot and n service nodes (or tasks), and that their locations are given. It is required to find m routes for vehicles to give prescribed services, with total length of routes as small as possible. The problems may include constraints such as capacity limitation, route length restriction, etc. Therefore there are many versions of vehicle routing problems, according to different purposes of practical applications.

For vehicle routing problems, research works have mainly been heuristics or approximation algorithms with theoretical supports or computational experiments, and mathematical modelling for applications.

This paper is a short presentation of our works. First, to delivery problems with capacity canstraints (see [2] or [1]), the eligibility of the penalty algorithm of Stewart & Golden[2] is proved, and an estimate for penalty gap, i.e. the difference between objective value of penalty solution and optimal solution, is obtained(see [3]). Secondly, a new heuristic is described for pick-up and drop-off problems(see [4]).

----

* Supported by the National Science Foundation of China.

§2. On Delivery Problems

Delivery problems with capacity constraints can be given in the following formulation P:

$$\text{Min} \sum_{k=1}^{m} \sum_{i,j=0}^{n} c_{i,j}\, x_{i,j,k} \tag{1}$$

$$\text{s.t.} \quad \sum_{i,j=0}^{n} d_i\, x_{i,j,k} \geqslant \sum_{i,j=0}^{n} d_i\, x_{i,j,k+1} \quad (k=1,2,\ldots,m-1) \tag{2}$$

$$\sum_{i,j=0}^{n} d_i\, x_{i,j,1} \leqslant D \tag{3}$$

$$x = (x_{i,j,k}) \in S^* \tag{4}$$

where m is number of vehicles, n is number of service nodes, $c_{i,j}$ is the distance from node i to node j, node 0 represents the depot, d_i is the demand at node i (d_0=0), D is the capacity of vehicles, the binary variable $x_{i,j,k}$ =1 when vehicle k goes to node j immediately after being at node i, otherwise $x_{i,j,k}$ = 0, and S^* represents the set of multi-traveling salesman solutions with less than m cycles all passing through node 0.

Stewart and Golden[2] suggested penalty problems P(λ) as follows:

$$\text{Min} \sum_{k=1}^{m} \sum_{i,j=0}^{n} c_{i,j} x_{i,j,k} + \lambda \sum_{i,j=1}^{n} d_i x_{i,j,1} \tag{5}$$

s.t. (2) & (4),

where λ is called penalty parameter. Problem P(λ) is similar to MTSP (multi-salesman travelling problem), and the heuristics for MTSP, such as arc exchange heuristics, can be adopted to solve P(λ). Now the penalty algorithm of Stewart and Golden[2] is composed of following steps:

Step1 $\lambda=0$, and solve P(0). If the solution x(0) satisfies (3), then x(0) is an optimal solution of problem P, and terminate. Otherwise, prescribe a step-length δ for λ, and go to Step 2.

Step 2 Set $\lambda=\lambda+\delta$, go to Step 3.

Step 3 Solve problem $P(\lambda)$. If the solution $x(\lambda)$ satisfies capacity constraints, then go to Step 4. Otherwise, go to Step 2.

Step 4 Search for smallest value $\lambda_* \in (\lambda-\delta, \lambda]$ such that the solution $x(\lambda_*)$ satisfies (3). Terminate, and take $x(\lambda_*)$ as an approximate solution.

The following result claims existence of λ_*, that is, the eligibility of the penalty algorithm. It is related to the combinatorial nature of the delivery problems, and usually it is not true for continuous cases.

Theorem 1 If the delivery problem P, i.e. (1)-(4), is feasible, then there exists λ_*, such that for any $\lambda \in [\lambda_*, \infty)$, the optimal solution $x(\lambda)$ of $P(\lambda)$ is feasible to P, and for any $\lambda \in [0, \lambda_*)$, $x(\lambda)$ is infeasible to P.

Let c(x) be the objective function of (1), d(x) be left side of (3), and x_* be optimal solution of P. The following theorem gives an estimate for penalty gap:

Theorem 2 $c(x(\lambda_*)) - c(x_*) \leqslant \lambda_*(D-d(x(\lambda_*)))$

§3. On Pick-up & Drop-off Problems

In pick-up & drop-off problems, each service node is replaced by a service task which includes one pick-up node and another drop-off node, so in the objective function, $c_{i,j}$ is generally asymmetric. Also the capacity constraints in delivery problems is replaced by route-length canstraints.

A new heuristic based on TSP value estimate, for the problem with rectangular distance and with a random depot and s random task nodes) is obtained from computer simulation. According to computational experiments, the TSP value is approximately equal to the product coefficient and the average distance between the depot and tasks, between all tasks. The coefficient is decided by simulation, and is slightly changable on s and shape of the rectangular region.

The heuristic given in [4] includes mainly steps of route expanse. The TSP value estimate is used to check the route-length constraint and to expand the route for making loading mileage ratio as large as possible. From the results of computational experiment, the new heuristic is comparable to and often better than the existing heuristics such nearest neighbour process and maximal saving process (see [1]).

References

[1] L.Bodin, B.Golden, A.Assad & M.Ball, Routing and Scheduling of Vehicle and Crews, the State of the Art, Computers & Operations Rese V.10, No.2(1983).

[2] W.Stewart & B.Golden, A Vehicle Routing Algorithm Based on Generalized Langrange Multipliers, Proc. of AIDS 1979 Annual Convention (Edited by L.Moore, etc.), V.2, New Orleans(1979), 108-110.

[3] W.Yu & Y.Fan, On the Penalty Algorithm for Delivery Problems In Chinese), Journal of Qufu Normal University, V.14, No.3(1988), 108-113.

[4] W.Yu & Y.Zhang, A New Heuristic for Pick-up & Drop-off Problems (in Chinese),Journal of Qufu Normal University, V.14, No.3(1988),120-125.